ENCYCLOPAEDIA OF OSCILLATIONS AND WAVES

ENCYCLOPAEDIA OF OSCILLATIONS AND WAVES

Vol. 1

Manpal Singh

ANMOL PUBLICATIONS PVT. LTD.
NEW DELHI - 110 002 (INDIA)

ANMOL PUBLICATIONS PVT. LTD.
H.O.: 4374/4B, Ansari Road, Darya Ganj,
New Delhi-110 002 (India)
Ph.: 23278000, 23261597
B.O.: No. 1015, Ist Main Road, BSK IIIrd Stage
IIIrd Phase, IIIrd Block,
Bangalore - 560 085 (India)
Visit us at: www.anmolpublications.com

Encyclopaedia of Oscillations and Waves

First Published, 2008

ISBN 978-81-261-3541-7 (Set)

PRINTED IN INDIA

Printed at Mehra Offset Press, Delhi.

Contents

Preface vii

1. **Superposition of Harmonic Waves** 1
Interference and Beats • Interference of Sound Waves • Interference of Two Waves • Young's Double-slit Experiment • Interference and Beats • Beats and Amplitude Modulation

2. **Electromagnetic Waves** 23
Ampere's Law • Maxwell's Equations

3. **Normal Modes of Vibration** 43
Coupled Oscillations

4. **Ultrasonics** 55
Audible, Infrasonic and Ultrasonic Waves • Ultrasonic Waves • Bats Avoiding Obstacles

5. **Acoustic Impedance** 67
Reflection and Refraction of Waves • Characteristic Impedance of a String • Acoustic Impedance

6. **Vibrations of Strings** 85
Rectangular Membranes • Equation of Motion • Laws of Vibration of Strings • Melde's Experiment

7. **Waves in Dispersive and Absorptive Media** 117
Wave Velocity • Group Velocity of Gravity Waves • Group Velocity of Surface Waves • Group Velocity of Matter Waves

8. **Velocity of Acoustic Waves** 135
Velocity of Longitudinal Waves • Velocity Sound • Effect of Temperature • Hebb's Telephone Method

9. **Stationary Waves** 149
Standing Waves • Waves in a Linear Bounded Medium

10. **Wave Motion** 165
Pressure and Energy • Transverse and Longitudinal Waves • Propagation of Sound Waves • Particle Velocity in Wave Motion

11. **Acoustics of Buildings** 203
Characteristic Features • Sabine's Reverberation Formula

12. **Recording and Reproduction of Sound** 219
Recording of Sound on Films • Magnetic Sound Recording

13. **Elementary Theory of Filters** 227
Electrical Wave Filters • Acoustic Filters

14. **Fourier's Theorem** 239
Periodic Functions • Significance of Fourier's Theorem

15. **Doppler's Effect** 269
Practical Application

Bibliography 293

Index 297

Preface

Simple Harmonic Oscillation is defined as the oscillation of a moving body, which is repeated after a definite interval of time. In oscillation, the moving body has the maximum velocity at the centre point and minimum velocity on the end points of the moving path. The moving bodies have velocity, acceleration, displacement amplitude and periodic time. Acceleration of the body is defined as the change in velocity per unit time. Displacement is defined as the distance of the moving body at any time from the centre point of the moving path. Amplitude of the moving body is defined as the maximum displacement of that body, on one side of the moving path, from its centre point. The acceleration is directly proportional to the displacement, but always directed towards the centre point. The time taken by the moving body in one complete oscillation is defined as the periodic time of the body. The periodic time depends upon effective length of the thread, from which the oscillatory body is hanged and on gravitational acceleration. In case of spring motion, the periodic time depends on the constant force of the spring and mass of the body.

When an object is set into oscillations, without any external influence, its amplitude of oscillation gradually decreases to zero. This is because the oscillating object is always subject to frictional force, arising from air resistance or from within the object itself. The force applied against motion is dissipated in form of heat out of the object only. When an oscillation body is displaced from its equilibrium-position and released, it begins to oscillate. The frequency of oscillations depends upon certain properties of the body. Such oscillations are called free oscillations and their frequency is called 'natural frequency' of the body. When a body is set into oscillation by an external periodic force of the same frequency as the natural frequency of the body, the amplitude of oscillation is increased considerably. This phenomena is called resonance. Hence the resonance is a special process of forced oscillation.

When the frequency of the external force is equal to the natural frequency of the body, then at each step the force helps the body in going ahead in its oscillation. The successive impulse given by the periodic force is added up and increase the amplitude of oscillation of the body continuously. The loss of energy of the body is due to air resistance and internal friction and it also increases with the increase of its amplitude of oscillation. When two harmonic wave-trains of the same frequency, having a constant phase relation travel simultaneously through a medium, the resultant wave intensity at any point is different from the sum of the intensities caused by separate wave-trains. This is the phenomenon of interference. When two harmonic wave-trains of the same frequency travel along the same line in opposite directions, their super position gives a pattern having alternately nodal points of zero displacement along with additional points of maximum displacement. This is known as 'stationary-wave pattern'. A practical source of light has a huge number of atoms, which emit discontinuously the trains of light waves at short intervals. These intervals are different for different atoms. Therefore, the phase of the light wave-train emitted by the innumerable atoms in the source changes with time in a random manner.

This exclusive work has a perfect account of harmonic oscillations in an interesting manner. Let's hope, this work would be able to serve its purpose by being beneficial for scientists, professionals, students and of course the general readers.

Editor

1

Superposition of Harmonic Waves

Interference and Beats

Principle of Superposition of Waves: This principle states that "when two (or more) waves travel simultaneously in an elastic medium, the resultant displacement of any particle at any instant is equal to the vector sum of the displacements of that particle corresponding to the separate waves at that instant.

The principle of superposition means that of the many waves, each one moves independently as if the others were not present at all; and that their individual shapes and other characteristics are not changed due to the presence of one another. This is seen in daily life.

When we listen to an orchestra, we receive a complex sound due to the superposition of many sound waves of different characteristics produced by different musical instruments. Still we can recognise the separate sounds of different instruments. Our radio antenna is open to the waves of different frequencies transmitted simultaneously by the different radio-stations. But when we adjust the radio to a particular station, we receive its programme as if the other stations were silent.

Derivation of the Principle of Superposition: Suppose y_1 is the displacement of a point x at time t due to a wave described by a function $y_1(x, t)$. It must be a solution of the general differential equation of wave motion, $\frac{\partial^2 y}{\partial x^2} = \frac{1}{v^2} \frac{\partial^2 y}{\partial t^2}$. That is

$$\frac{\partial^2}{\partial x^2} y_1(x,t) = \frac{1}{v^2} \frac{\partial^2 y}{\partial t^2} y_1(x,t) \qquad \text{...(i)}$$

Similarly, let y_2 be the displacement of the same point x at the same time t due to another wave described by $y_2(x, t)$. This must also be a solution of the differential equation. That is,

$$\frac{\partial^2}{\partial x^2} y_2(x,t) = \frac{1}{v^2} \frac{\partial^2}{\partial t^2} y_2(x,t). \qquad \text{...(ii)}$$

Adding eq. (i) and (ii), we get

$$\frac{\partial^2}{\partial x^2} \{y_1(x, t)+y_2(x, t)\} = \frac{1}{v^2} \frac{\partial^2}{\partial t^2} \{y_1(x, t)+y_2(x, t)\}$$

or

$$\frac{\partial^2}{\partial x^2} y(x, t) = \frac{1}{v^2} \frac{\partial^2}{\partial t^2} y(x, t), \qquad \text{...(iii)}$$

where $y(x, t)$ is the sum of the two functions:

$$y(x, t) = y_1(x, t) + y_2(x, t). \qquad \text{...(iv)}$$

Eq (iii) shows that the sum of the two wave functions described by eq. (iv) also satisfies the general differential equation as each separate wave function does. Hence $y(x, t)$ is *a* proper function to describe the displacement of the point x at time t. Eq. (iv) shows that the displacement y of the point x at time t is equal to the sum of the displacements y_1 and y_2 of that point due to the separate waves at the same time. Thus the principle of superposition is a consequence of the *form* of the differential equation of wave motion.

Limitation: The principle of superposition holds for those waves only whose equations of motion are *linear (i.e.* obey Hooke's law). Thus it does not hold for shock waves (created by explosions) and water-waves.

Phenomena arising from Superposition of Waves: The superposition of waves gives rise to various phenomena like interference, diffraction, beats and stationary waves.

Interference and Diffraction: When two harmonic wave-trains of the same frequency and having a constant phase relation travel simultaneously through a medium, the resultant wave intensity at any point is different from the sum of the intensities due to the separate wave-trains.

This is the phenomenon of "interference". It arises because at each point of observation the phase difference between the superposing waves depend upon the different paths travelled by the separate waves and the resulting amplitude (and hence intensity) may be greater or less than that of any single wave. Thus there is a distribution of intensity in space. This distribution is known as "interference pattern".

When an advancing wave is obstructed by some barrier then each unobstructed point on the original wave front acts as a new source, and the disturbance beyond the barrier is the superposition of all the waves spreading out from these new secondary sources As a result there is a characteristic intensity distribution in the region beyond the barrier. This distribution is known as "diffraction pattern".

Beats: The principle of superposition results in another type of interference, which we may call interference in time It occurs when two wave-trains of slightly different frequencies travel through the same region. Then the amplitude of the resultant wave at any point is not constant but varies in time. In case of sound waves the varying amplitude gives rise to variations in loudness which are called 'beats'.

Stationary (or Standing) Waves: When two harmonic wave-trains of the same frequency travel along the same line in opposite directions, their superposition gives a pattern having alternately nodal points of zero displacement and antinodal points of maximum displacement. This is known as 'stationary-wave pattern' having no energy-transmission.

Interference of Sound Waves

When two wave-trains of the same frequency having a constant phase relation travel simultaneously through a medium, the resultant wave intensity at any point is different from the sum of the intensities due to the separate wave-trains. The wave trains are then said to have interfered with each other, and the phenomenon is called as "interference."

Let S_1 and S_2 be two point-sources emitting sound waves of the same amplitude and same frequency.

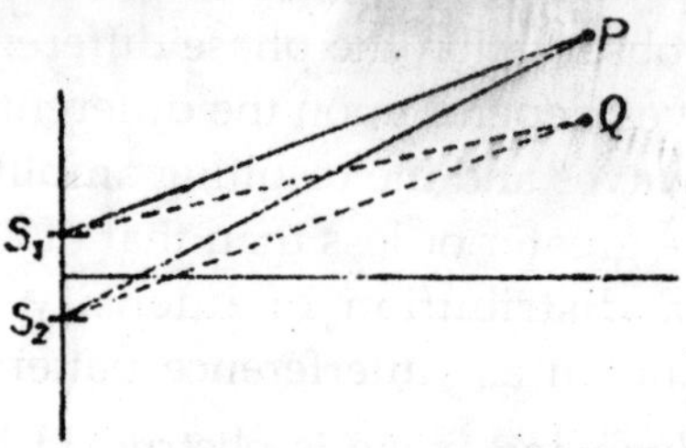

Figure

If at any point *P*, the two waves reach in the same phase, *i.e.* the compression (or rarefaction) of one wave coincides with the compression (or rarefaction) of the other, the resultant amplitude at *P* is equal to the sum of the amplitudes due to the separate waves. Hence the resultant intensity of sound is a maximum.

The interference at *P* is said to be "constructive". It at any other point *Q* the waves reach in opposite phase, *i.e.* the compression of one wave coincides with the rarefaction of the other, the resultant amplitude at *Q* is equal to the difference of amplitudes of the two waves. Hence the resultant intensity is a minimum. The interference at *Q* is said to be "destructive." Thus there is a redistribution of energy in the phenomenon of interference.

Superposition of two Plane Harmonic Waves of same Period: Let us consider two simple harmonic waves of the same frequency travelling in the same direction. Let a_t and a_2 be the amplitudes of the waves and ϕ the phase difference between them at any point in the medium. If the displacements due to these waves at that point be y_1, and y_2, then,

$$y_1 = a_1 \sin \omega t \quad \text{...(i)}$$

and $$y_2 = a_2 \sin (\omega t + \phi). \quad \text{..(ii)}$$

By the principle of superposition, the resultant displacement at that point is given by

$$\begin{aligned} y &= y_1 + y_2 \\ &= a_1 \sin \omega t + a_2 \sin (\omega t + \phi) \\ &= a_1 \sin \omega t + a_2 \sin \omega t \cos \phi + a_2 \cos \omega t \sin \phi \\ &= \sin \omega t \, (a_1 + a_2 \cos \phi) + \cos \omega t \, (a_2 \sin \phi). \end{aligned}$$

Let $$a_1 + a_2 \cos \phi = R \cos \theta \quad \text{...(iii)}$$

and $$a_2 \sin \phi = R \cos \theta, \quad \text{...(iii)}$$

where R and θ are new constants. Then

$$\begin{aligned} y &= \sin \omega t \, R \cos \theta + \cos \omega t \, R \sin \theta \\ &= R \sin (\omega t + \theta). \end{aligned}$$

This equation is similar to eq. (i) and (ii). Hence the resultant displacement at that point is due to a resultant wave of the same type as the interfering waves, but having an amplitude R. To determine R, we square eq. (iii) and (iv) and then add:

$$R^2 \cos^2 \theta + R^2 \sin^2 \theta = (a_1 + a_2 \cos \phi)^2 + (a_2 \sin \phi)^2$$

or $$R^2 = a_1^2 + a_2^2 + 2a_1 a_2 \cos \phi.$$

The wave intensity I, being directly proportional to the square of the amplitude, is given by

$$I = a_1^2 + a_2^2 + 2a_1 a_2 \cos \phi. \quad \text{...(v)}$$

Thus the resultant intensity at any point depends upon the phase difference ϕ between the two waves at that point.

Maximum Intensity: The intensity I will be maximum at points for which $\cos \phi = +1$, that is, phase difference, $\phi = 0, 2\pi, 4\pi, \ldots\ldots$

$$= 2m\pi \qquad \text{where } m = 0, 1, 2, \ldots$$

Then from eq. (v), we have

$$I_{max} = a_1^2 + a_2^2 + 2a_1a^2 = (a_1 + a_2)^2.$$

Thus at points where the two interfering waves meet in the same phase ($\phi = 0, 2\pi, 4\pi \ldots$), the resultant intensity $(a_1+a_2)^2$ is greater than the sum of the intensities $(a_1^2+a_2^2)$ of the two waves, and the waves are said to interfere constructively.

Minimum Intensity: The intensity I will be minimum at points for which $\cos\phi = -1$, that is,

phase difference, $\phi = \pi, 3\pi, 5\pi \ldots\ldots$

$$= (2m+1)\pi, \qquad \text{where m} = 0, 1, 2, \ldots$$

Then from eq. (v), we have

$$I_{min} = a_1^2 + a_2^2 - 2a_1a^2 = (a_1 - a_2)^2.$$

Thus at points where the two interfering waves meet in opposite phase ($\phi = \pi, 3\pi, 5\pi, \ldots$), the resultant intensity (a_1-a_2) is less than the sum of the intensities $(a_1^2+a_2^2)$ of the two waves, and the waves are said to interfere destructively.

At intermediate points the intensity is found to be between its maximum and minimum values.

Thus if we move over the space of overlap of the two waves we shall pass alternately through the regions of intensity maxima and minima. The pattern formed by these regions is called the "interference pattern."

Conditions for Interference: For interference between the two waves to occur, the following conditions must be fulfilled:

(i) *The Phase Difference between the Interfering Waves should Remain Constant:* If the phase difference between the two waves at a point varies with time, then the intensity of sound at that point will also vary with time and the interference will not be sustained. Hence, for observing the interference, both the waves should originate from the same source.

(2) *The Amplitudes of the Two Waves Should be Nearly Equal:* If the amplitudes differ appreciably then there will be sufficient intensity even at points where the waves meet in opposite phase. Hence the interference will not be clear. For best interference, the amplitudes of the waves should be exactly equal.

(3) *The Displacements Produced by the Two Waves should be Along the Same Line:* It is not so then the displacement at a point where the waves meet in opposite phase will not be zero.

Interference of Two Waves

When two waves travel simultaneously through a medium, each wave proceeds as though the other were not present. In this sense, the two waves *do not* change the progress of each other's motion. However, at each point in the medium, the total displacement is just the sum of the displacements corresponding to the separate waves.

At points where the two waves arrive in phase with each other, the wave intensity is a maximum; at points where the two waves arrive in opposite phase, the wave intensity is a minimum. This phenomenon is "interference" and it does not change the progress of either wave.

Interference and Conservation of Energy: In the interference phenomenon the energy of the medium due to passage of the two waves remains conserved, it is simply redistributed over the interference pattern. This can be proved mathematically :

The resultant intensity at a point in the interference pattern produced by two waves of amplitudes a_1 and a_2 can be written as

$$I = a_1^2 + a_2^2 + 2a_1a_2 \cos\phi,$$

where ϕ is the phase difference between the waves at that point. The average intensity between the range $\phi=0$ to $\phi=2\pi$ is

$$\bar{I} = \frac{\int_0^{2\pi} I d\phi}{\int_0^{2\pi} d\phi} = \frac{\int_0^{2\pi} (a_1^2 + a_2^2 + 2a_1\, a_2 \cos\phi) d\phi}{\int_0^{2\pi} d\phi}$$

$$\frac{\left[a_1^2\phi + a_2^2\phi + 2a_1 a_2 \sin\phi\right]_0^{2\pi}}{\left[\ \phi\ \right]_0^{2\pi}}$$

$$= \frac{2\pi(a_1^2 + a_1^2)}{2\pi}$$

$$= a_1^2 + a_2^2.$$

Thus the average intensity is equal to the sum of the separate intensities. That is, whatever energy apparently disappears at the minima is actually present at the maxima. Thus there is no loss of energy in the interference phenomenon.

Essential Conditions for Interference Pattern

Coherent Sources: Two "point" sources having the same frequency have a phase difference between them which does not change with time. Such sources are said to be "coherent". When two coherent sources are located at different points, the waves emitted from them maintain a constant phase difference at a given point in space in the region of overlap.

Hence a point that is one of maximum intensity at a given time will always be a point of maximum intensity, and likewise a point of minimum intensity will always be a point of minimum intensity. In other words, a stationary interference pattern will be observed. This 'coherency' between the two sources is an essential condition for obtaining interference pattern.

Interference between Waves from Independent Sources: In practice we do not have "point" sources. A practical source of light has a huge number of atoms which emit discontinuously trains of light waves at short intervals. These intervals are different for different atoms. Therefore the phase of the light wave-train emitted by the innumerable atoms in the source changes with time in a random way.

Hence the waves from two 'independent' light sources will meet at a point with a phase difference changing with time. Therefore, if at a certain instant the two waves interfere at that point constructively producing maximum intensity, a short time later they will interfere at the same point destructively producing minimum intensity. Thus the two waves will produce a time-averaged uniform intensity throughout, i.e. no interference pattern would be observed.

In the case of sound waves, the sources producing them are sometimes (but rarely) "single", and they can be matched in frequency more or less exactly. Therefore, interference between waves from separate sources, like two tuning forks, is observable. However, waves from two violins cannot give a sustained interference effect because a violin is not a single source.

In practice, interference effects are obtained by means of two coherent sources obtained by special means. Two vertical rods touching the surface of water and driven by a common device produce an interference pattern on the water surface.

Two parallel slits receiving light from a common narrow illuminated slit produce an interference pattern of light on a screen. Sound waves emitted by two loudspeakers driven by a common audio-oscillator produce an interference pattern of sound.

In all these cases any change in the phase of the common 'driver' is transmitted simultaneously to the two sources, so that at any point a constant phase difference is maintained between the two waves.

Young's Double-slit Experiment

One of the best known methods for producing optical interference effects is the Young's double-slit experiment. Here the two coherent sources are two parallel narrow slits S_1 and S_2 very close to each other and illuminated by a monochromatic wave-train from a single source S equidistant from S_1 and S_2.

The light waves from S reach the slits S_1 and S_2 in the same phase. According to Huygen's principle, S_1 and S_2 act as new light-sources from which secondary wavelets originate. These waves form interference fringes on a distant screen. Let us calculate the width of these fringes.

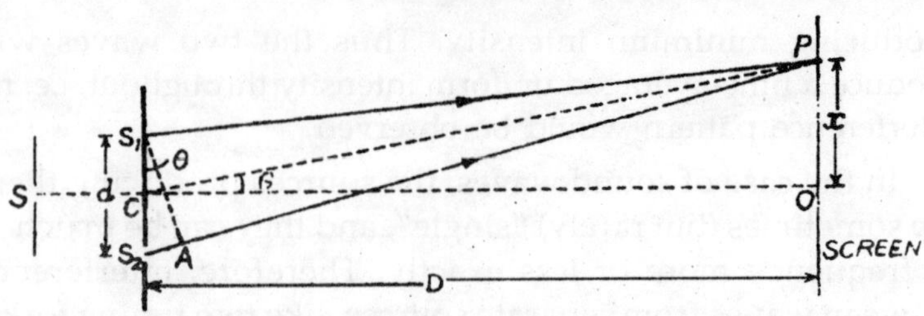

Figure

Let d be the separation between the slits S_1 and S_2, and D the distance of the screen from the plane of the slits. Let the right-bisector CO of the double-slit S_1S_2 meet the screen at O. Let P be an observation point, distant x from 0. The waves from S_1 and S_2 travel different distances S_1P and S_2P to reach P. Let S_1A be perpendicular drawn on S_2P from S_1. Then the path difference between the two waves at the point P is S_2A.

In similar triangles S_1S_2A and PCO, we have

$$\frac{S_2A}{S_1S_2} = \frac{OP}{CP} \simeq \frac{OP}{CO}$$

Since $CO >> S_1S_2$; we may take $CP \simeq CO$. Putting the values

$$\frac{S_2A}{d} = \frac{x}{D}$$

or path difference, $S_2A = \dfrac{xd}{D}$

Now, the intensity at the point P is the maximum or minimum according as the path difference S_2A is an integral multiple of wavelength or an odd multiple of half-wavelength.

Thus for maximum intensity (bright fringe), we have

$$\frac{xd}{D} = m\lambda \qquad \text{where } m = 0,1,2........$$

or $$x = m\frac{D\lambda}{d},$$

and for minimum intensity (dark fringe), we have

$$\frac{xd}{D} = (2m+1)\ \frac{\lambda}{2} = \left(m+\frac{1}{2}\right)\frac{\lambda}{2} \text{ where } m = 0,1,2,\dots\dots$$

or $$x = (m+\tfrac{1}{2})\ \frac{D\lambda}{d}.$$

Now, let x_m and x_{m+1} be the distances of the *m*th and (m+l)th bright fringes. Then

$$x_m = m\frac{D\lambda}{d}$$

$$x_{m+1} = (m+1)\frac{D\lambda}{d}$$

$\therefore$ spacing between *m*th and (*m*+l)th bright fringes is

$$x_{m+1} - x_m\ (m+1)\frac{D\lambda}{d} - m\frac{D\lambda}{d} = \frac{D\lambda}{d}.$$

It is independent of *m*. Thus the spacing between any two consecutive bright fringes is the same. Similarly, it can be shown that the spacing between two dark fringes is also $D\lambda/d$.

The spacing between two consecutive bright or dark fringes is called 'fringe-width' *W*. Thus

$$W = \frac{D\lambda}{d}.$$

This is the required expression. It shows that the fringe-width is inversely proportional to the separation between the sources and directly proportional to the wavelength.

Beats

When two sound sources of *slightly different* frequencies are sounded simultaneously, the intensity of the resultant sound rises and falls alternately with time.

This phenomenon is called "beats". It arises as a result of the interference between the sound wave-trains emitted from the two sources.

Explanation: Suppose at a certain instant the two waves are meeting in the same phase at a particular point. They will produce maximum sound intensity. But the two waves are of *slightly different* frequencies. Therefore, from this instant, they will get further and further out of step until one gains half an oscillation on the other. Then, at the same point, the two waves will meet in opposite phase and produce minimum sound intensity. Again, after some time one wave will gain one full oscillation on the other.

Then they will once again meet in phase and produce maximum sound, and so on. If the difference in frequencies of the two sources is *n,* then in 1 second one wave will gain *n* oscillations over the other. Hence the intensity of sound will rise and fall 'n' times per second, giving 'n' beats per second. Thus the number of beats per second is equal to the difference in frequencies of the two sources.

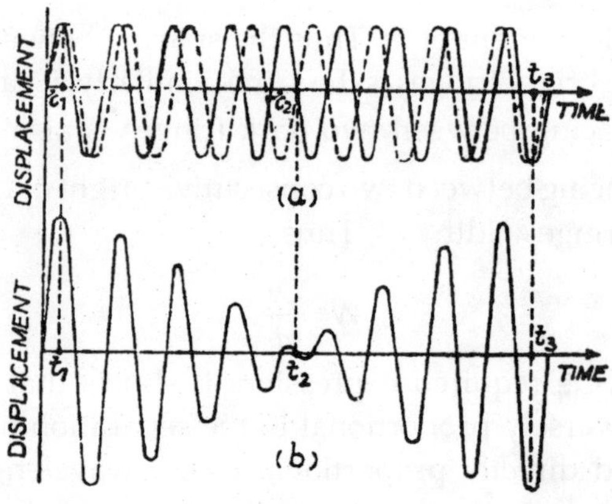

Figure

Graphical Representation: In figure above are plotted the displacements at a point produced by the two waves separately against time. The resultant displacement at that point is the sum of the individual displacements and is plotted in figure. At time t_1, the two displacements are in phase and add to each other. But, on account of difference in frequencies, they gradually get more and more out of step until at time t_2 they practically cancel each other. Again, they get more and more in step until once more at t_2 they reinforce each other. The resultant curve clearly shows the variation in amplitude and hence in intensity.

Experimental Demonstration: The production of beats can be demonstrated by having two tuning forks of same frequency. When they are sounded together, there are no beats. If a small piece of wax is fixed to a prong of one of them, the frequency of that fork will be reduced slightly and then, when it is sounded with the first fork, beats will be heard.

Mathematical Investigation: Let two waves having same amplitude a, but "slightly different" frequencies n_1 and n_2 be travelling simultaneously in a medium. Let us represent the displacement at a point produced by one wave as

$$y_1 = a \sin 2\pi n_1 t$$

and the displacement produced by the other wave as

$$y_2 = a \sin 2\pi n_2 t.$$

The resultant displacement at the point is

$$y = y_1 + y_2 = a\ (\sin 2\pi n_1 t + \sin 2\pi n_2 t)$$

Applying the theorem $\sin A + \sin B = 2 \sin \frac{A+B}{2} \cos \frac{A-B}{2}$, we get

$$y = 2a \sin\left\{2\pi\left(\frac{n_1+n_1}{2}\right)t\right\} \cos\left\{2\pi\left(\frac{n_1-n_1}{2}\right)t\right\}$$

$$= 2a\left\{\cos 2\pi\left(\frac{n_1-n_1}{2}\right)t\right\} \sin\left\{2\pi\left(\frac{n_1+n_1}{2}\right)t\right\}$$

where $A = 2a\left\{\cos 2\pi\left(\frac{n_1-n_1}{2}\right)t\right\}$. Thus the resultant vibration

of the point has a frequency $\frac{n_1+n_2}{2}$ and amplitude A; which is given by

$$A = 2a \cos\left\{2\pi\left(\frac{n_1-n_1}{2}\right)t\right\}$$

This shows that A is not constant but periodically varies

with time between a maximum value $2a$ and a minimum value 0. Hence the intensity of the resultant sound, being proportional to the square of the amplitude, rises and falls alternately.

Now, the amplitude A is maximum when

$$\cos\left\{2\pi\left(\frac{n_1 - n_1}{2}\right)t\right\} = \pm 1$$

$$2\pi\left(\frac{n_1 - n_2}{2}\right) t = k\pi, \text{ where } k=0, 1,2,3,...$$

$$\therefore \qquad t = \frac{k}{n_1 - n_2}$$

$$0 = \frac{1}{n_1 - n_2}, \frac{2}{n_1 - n_2}, \frac{3}{n_1 - n_2}, \text{ etc.}$$

$\therefore$ amplitude maxima occur at intervals of $\frac{1}{n_1 - n_2}$ second, or $(n_1 - n_2)$ times per second.

Similarly, the amplitude is minimum when

$$\cos\left\{2\pi\left(\frac{n_1 - n_1}{2}\right)t\right\} = 0$$

or $$2\pi\left(\frac{n_1 - n_2}{2}\right) t = (2k + 1)\frac{\pi}{2}, \qquad \text{where } k=0, 1,2,3, \text{etc.}$$

$$\therefore \qquad t = \frac{(2k+1)}{2(n_1 - n_2)}$$

$$= \frac{1}{2(n_1 - n_2)}, \frac{3}{2(n_1 - n_2)}, \frac{5}{2(n_1 - n_2)}, \text{ etc.}$$

$\therefore$ amplitude minima occur between the maxima, at intervals of $\frac{1}{n_1 - n_2}$ second or $(n_1 - n_2)$ times per second. Thus beat frequency$=n_1 - n_2$.

Interference and Beats

When two interfering sound waves have exactly the same frequency, then the phase difference between them at any point remains constant. Therefore, the intensity of the resultant sound at that point remains constant. Hence the alternate regions of loudness and silence which are formed remain stationary. In other words, the interference pattern remains fixed in space. This is a case of stationary interference in space, and called simply as "interference."

If, however, the two waves slightly differ in frequencies, the phase difference at any point varies with time. If at a certain instant the two waves are in phase at a particular point, then after a short time when one wave has gained half a vibration over the other, they will become in opposite phase, and so on. Thus they will produce maximum and minimum intensity alternately at the same point. This is a case of interference in time and is known as "beats".

Applications of Beats: There are a number of applications of the phenomenon of beats:

(i) *Determination of Frequency:* The phenomenon of beats can be used to determine the frequency of a tuning fork. The given tuning fork, say A, and an another tuning fork B whose frequency is known are sounded together. The number of beats produced in a measured interval of time is counted. If n be the frequency of B and x the number of beats produced per second, then the frequency of A must be

$$\text{either } n+x \text{ or } n-x.$$

Now, one prong of the unknown fork A is loaded with a little wax and the two forks are sounded again. If the number of beats per second is decreased then the frequency of the fork A must be $n+x$, but if the number of beats per second is increased, then the frequency of A must be $n-x$.

(The frequency of the fork A is lowered when it is loaded with wax. Thus if its frequency was already greater than that of B, then on loading, the difference between the frequencies of A and B must decrease and hence the number of beats per second must alto decrease. On the other hand, if the frequency of A was smaller than that of B, the on loading A, the difference between the frequencies of A and B must increase. Hence the number of beats per second must also increase.)

(ii) *Toning of Musical Instruments*: The musical instrument is sounded together with another with which it is to be tuned.

If the frequencies are slightly unequal, beats are heard. Now the frequency of the given instrument is altered until the beats disappear. This is called 'tuning'.

(iii) *Detection of harmful Gases in Mines:* The apparatus used to detect harmful gases in mines has two similar small pipes, one filled by pure air and the other by mine-air. These pipes are blown together. When the mine-air is also pure, the pipes emit sound of the same frequency and no beats are heard. When 'firedamp' is present, the density of the mine-air decreases and the velocity of sound in it is changed. Consequently, the frequency of the sound produced by the mine-air pipe is slightly changed and beats are heard. The miners have tables with them to estimate the percentage of 'fire-damp' from the number of beats heard. This gives warning before hand to the miners before the mine-air is dangerous enough to cause an explosion.

(iv) *In Radio Reception:* Beats are used in the heterodyne reception of radio waves. The incoming high-frequency signals are mixed with the waves of slightly different frequency generated at the receiving station. These two combine together to produce pulsations of frequency lying in the audible range.

Beats and Amplitude Modulation

When two waves of equal amplitudes a but frequencies n_1 and n_2 differing by a small amount superpose, they give rise to a wave-system of frequency $(n_1+n_2)/2$ but with an amplitude varying periodically with time between $2a$ and 0. The maximum and minimum amplitudes occur at a frequency $n_1 \sim n_2$. This is the phenomenon of "beats" in which the amplitude is said to be "modulated". In this example, the amplitude varies between $2a$ and zero ; this is complete or 100 per cent modulation.

Amplitude modulation is very common in radio transmission. The audio-frequency sound waves (speech or music), converted into electrical signal at the broadcasting station, cannot be transmitted as such to distant places because of two reasons : (i) Their intensity decreases rapidly with distance. (ii) An enormously large transmitting antenna will be required. Hence, in practice, a *radio-frequency* wave (produced by an oscillator) is used to carry the desired audio-frequency signal. The process of superposing' the audio-frequency signal on the *carrier* radio-frequency wave is called "modulation". In this process the amplitude of the carrier wave varies in accordance with the audio-frequency signal which is to be transmitted.

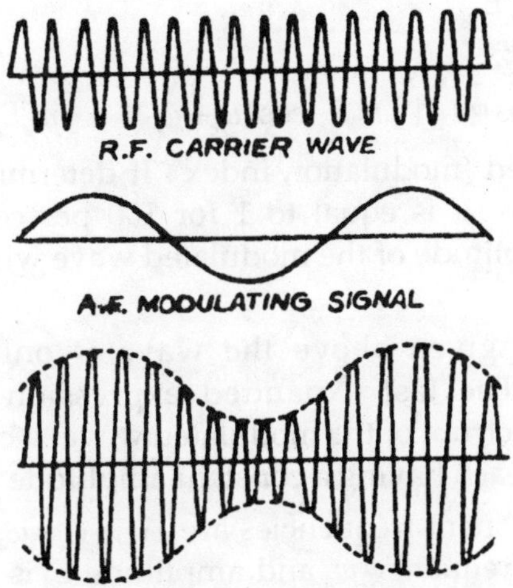

MODULATED WAVE

Figure given above shows the radio-frequency carrier wave, the audio-frequency modulating signal and the resultant modulated wave.

Let the carrier wave be represented as

$$e_c = E_m \cos \omega_c t.$$

and the modulating signal as

$$e_m = E_m \cos \omega_m t,$$

where $\omega_m << \omega_c$. Since the amplitude of the modulated carrier wave varies in accordance with the modulating signal; it should be of the form

$$E = E_c + K_a e_m$$

where K_a is a proportionality factor. Thus the modulated wave may be represented as

$$e = E \cos \omega_c t$$

$$= (E_c + K_a e_m) \cos \omega_c t$$

$$= (E_c + K_a E_m \cos \omega_m t) \cos \omega_c t$$

$$= E_c \cos \omega_c t + m_a E_a \cos \omega_m t \cos \omega_c t$$

where $m_a = \dfrac{K_a E_m}{E_c}$

$$= E_c \cos \omega_c t + \{ \tfrac{1}{2} m_a E_a \cos(\omega_c + \omega_m)t + \tfrac{1}{2} m_a E_a \cos(\omega_c + \omega_m)t \}.$$

m_a is called 'modulation index'. It determines the degree of modulation. It is equal to 1 for 100 per cent modulation when the amplitude of the modulated wave will vary between $2E_c$ and zero.

In figure given above the wave is only 50 per cent modulated. The last expanded expression indicates the frequency spectrum of the modulated wave. It consists of three components, each having a constant amplitude and frequency.

The component frequencies are ω_c, $(\omega_c + \omega_m)$ and $(\omega_c - \omega_m)$. The wave of frequency w_c and amplitude E_e is just the carrier wave, while the other two waves of frequencies $(\omega_c + \omega_m)$ and $(\omega_c - \omega_m)$ and same amplitude $\frac{1}{2} m_a E_c$ are known as 'upper sideband' and 'lower sideband' respectively. The generation of side bands, however, lead to the crowding of radio frequencies and interference between stations.

Problems

1. *In a Young's double-slit experiment interference fringes are formed using a monochromatic source of light of wavelength 5890 Å. The slits are 0.5 nun apart and the screen is at a distance of 1.50 metre from the slits. Calculate the fringe-width.*

Solution: The fringe-width is given by

$$W = \frac{D\lambda}{d}$$

Here, λ =5890 Å=5890×10^{-10} m, slit-separation d=0.5mm

$= 0.5 \times 10^{-3}$ m and screen distance D=1.50 m.

$$\therefore \quad W = \frac{1.50 \times (5890 \times 10^{-10})}{0.5 \times 10^{-3}} = 1.77 \times 10^{-3} \text{m}$$

$$= 1.77 \text{ mm.}$$

2. A sound wave of unknown frequency gives 10 beats with a wave of frequency 300 Hz and 15 beats with a wave of frequency 325 Hz. What is the frequency of the unknown wave?

Solution: The number of beats per second is equal to the difference in frequency of the two waves. Then, if n is the frequency of the unknown wave, we have

$$n = 300 \pm 10 = 310, 290$$

and $$n = 325 \pm 15 = 340, 310.$$

Obviously, the frequency of the unknown wave is 310 Hz.

3. Calculate the velocity of sound in a gas in which two waves of lengths 5.0 and 55 metre produce 6 beats in 1 second.

Solution: Let v metre/sec be the velocity of sound. Then the frequencies of the two waves will be

$$n_1 = \frac{v}{\lambda_1} = \frac{v}{5.0} \text{ and } n_2 = \frac{v}{\lambda_2} = \frac{v}{5.5}$$

As the number of beats per second is 6, we have

$$n_1 - n_2 = \frac{v}{5.0} - \frac{v}{5.5} = 6$$

$$v\left(\frac{5.5 - 5.0}{5.0 \times 5.5}\right) = 6$$

$$v = \frac{6 \times 27.5}{0.5} = 330 \text{ m/sec.}$$

4. A tuning fork of unknown frequency gives 5 beats when with another tuning fork of frequency 384 Hz. When the other fork is loaded with a little wax, only 3 beats are heard. Calculate the frequency of the fork.

Solution: The forks produce 5 beats in 1 second. Hence the difference between their frequencies is 5. As the frequency of the second fork is 384, that of the first should be

$$384 \pm 5 = 389 \text{ or } 379.$$

When the prong of the second fork is loaded, its frequency decreases slightly. But now the beat-frequency also decreases. Therefore, the frequency of the first fork must be less than 384. that is, it is 379.

5. *A fork of unknown frequency gives 4 beats per second when sounded with another, of frequency 256. The fork is now loaded with a piece of wax and again 4 beats per second are heard. Calculate the frequency of the unknown fork.*

Solution: The frequency of one fork is 256 and it gives 4 beats per second with unknown tuning fork. Therefore, the frequency of the unknown fork must be 256±4=260 or 252.

When the unknown fork is loaded with wax, its frequency decreases. If its frequency is taken to be 252, then on loading, the frequency would decrease and so the number of beats it would give with the fork of frequency 256 must increase.

If the frequency of the unknown fork is taken to be 260, then on loading it may decrease to 252 when it will again give 4 beats with the fork of frequency 256. Hence

frequency of unknown fork before loading=260.

6. *Calculate the number of beats heard per second if there are three sources of sound of frequencies 400, 401 and 402 of equal intensities sounded together.*

Solution: Let a be the amplitude, and $(n-1)$, n and $(n+1)$ the frequencies of the three waves. The equation for the resultant displacement is

$$y = a \sin 2\pi (n-1) t + a \sin 2\pi nt + a \sin 2\pi (n+1)t.$$

Adding the first and the third terms, we get

$$\begin{aligned} y &= a \sin 2\pi nt + 2a \sin (2\pi nt) \cos 2\pi t \\ &= a \sin 2\pi nt (1+2 \cos 2\pi t) \\ &= a (1+2 \cos 2\pi t) \sin 2\pi nt. \end{aligned}$$

This shows that the resultant amplitude is a (1+2 cos 2 π t). It is a maximum when

$$\cos 2\pi t = +1$$

or $\quad 2\pi t = 2k\pi,$ where k=0, 1, 2, 3...

or $\quad = k = 0, 1, 2, 3,$

∴ amplitude maxima occur at intervals of 1 second. Similarly, the resultant amplitude is minimum when

$$1 + 2\cos 2\pi t = 0$$

or $\quad \cos 2\pi t = -\frac{1}{2}$

or $\quad 2\pi = 2k\pi + \frac{2\pi}{3}$ where k=0, 1, 2,.

$$t = k + \frac{1}{3} = \frac{1}{3}, \frac{4}{3}, \frac{7}{3}, \frac{10}{3}, \ldots\ldots$$

∴ amplitude minima also occur at intervals of I second. Thus the resultant sound passes through one maximum and one minimum in each second. In other words, the number of beats per second is one.

2

Electromagnetic Waves

The electromagnetic waves consist of oscillating electric and magnetic fields, which are at right angles to each other and also at right angles to the direction of wave propagation. Thus, electromagnetic waves are transverse in nature. Unlike mechanical waves, the electromagnetic waves travel through a vacuum with the speed of light c. Furthermore, they carry energy and momentum and hence exert pressure on striking a surface.

Production of Electromagnetic Waves: The electromagnetic waves are generated by accelerating electric charges. For example, electrons moving from a higher to a lower energy level in an atom radiate an electromagnetic wave (light wave) of a particular wavelength.

A very hot ionised gas consisting of charged particles radiates electromagnetic waves over a continuous spectral range. A radio-station antenna emits electromagnetic waves (radio waves) due to the oscillating charges in the alternating current flowing in the antenna. After the theoretical prediction of their existence by Maxwell, the electromagnetic waves were first generated and detected by Hertz.

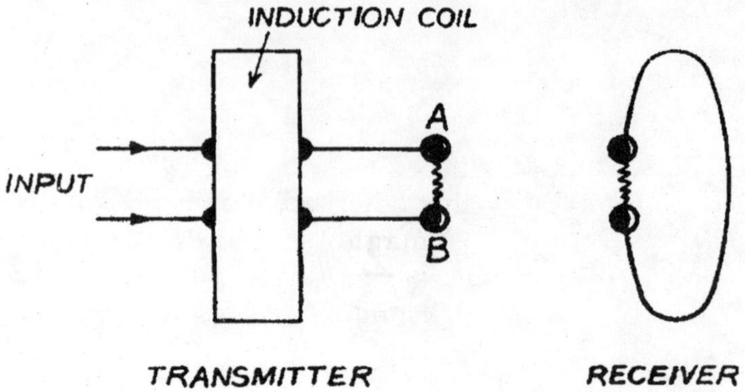

His apparatus is consists of two spherical electrodes *A* and *B* with a narrow air gap between them. These electrodes are connected to an induction coil which provides short voltage surges to the spheres, charging one positive and the other negative. When the voltage between the spheres reaches the breakdown voltage for air, a spark is generated in the air gap.

As the air in the gap is ionised, it conducts more readily and the discharge between the spheres becomes oscillatory. As a result of the oscillation (and hence acceleration) of free charges in the loop, electromagnetic waves are radiated from the air gap.

The air gap is thus a transmitter of electromagnetic waves. It is equivalent to an *L-C* circuit, the *L* is due to the loop and the *C* is due to the spheres.

A loop of wire having a small air gap is used as a detector (or receiver) of electromagnetic waves. This loop, placed several metres from the transmitter, has its own effective inductance, capacitance and natural frequency of oscillation.

Sparks were induced across the gap of the receiver when its frequency was adjusted to match the frequency of the transmitter. Thus Hertz showed that the oscillating current induced in the receiver was produced by electromagnetic waves radiated by the transmitter.

The Spectrum of Electromagnetic Waves: Hertz, in 1887, generated and detected the radio-frequency electromagnetic waves predicted by Maxwell. At that time only the radio waves and visible light were considered to be the electromagnetic waves.

It is now known that other radiation like X-rays, y-rays, etc. are also electromagnetic waves. All forms of electromagnetic waves exhibit the same physical characteristics and are distinguished by their wavelength (or frequency) only.

The range of wavelengths (or frequencies) of the electromagnetic waves is very wide and on this basis they can be given an order.

This order is called the 'electromagnetic spectrum'. It ranges from the very large radio waves to the very small y-rays. The visible light is only a very small part of the electromagnetic spectrum.

Figure given below shows the wavelength (or frequency) spectrum of electromagnetic waves. There is, however, no sharp dividing line between one form of wave and the next.

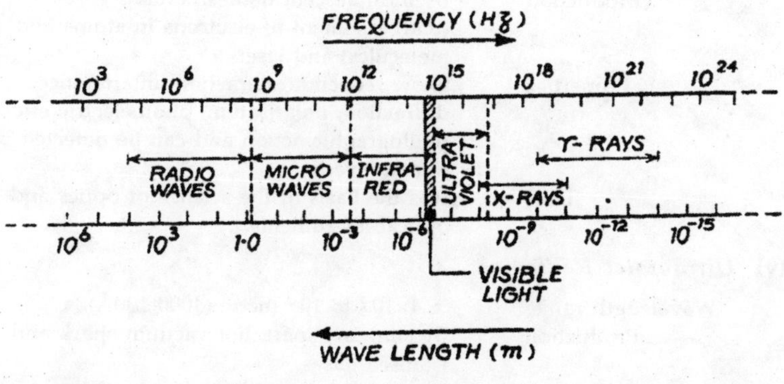

The wavelength-ranges, methods of production, important properties and uses of the various forms of electromagnetic waves are as follows:

(i) *Radio Wares:*

Wavelength range	:	$\approx 10^4$ to 1 metre.
Production	:	By electronic devices as oscillating electric *(L-C)* circuits.
Properties	:	They are reflected, refracted and diffracted.
Uses	:	In radio and television communication systems.

(ii) *Microwaves (Short Radio Waves):*

Wavelength range	:	≈ 1 to 10^{-3} metre.
Production	:	By spark discharge and electronic devices as magnetron and klystron.
Properties	:	They are reflected, refracted and diffracted, produce spark in the gap of receiving circuits.
Uses	:	In radar systems and for studying the atomic and molecular properties of matter.

(iii) *Infrared Waves:*

Wavelength range	:	$\approx 10^{-3}$ to 7.8×10^{-7} metre.
Production	:	By hot bodies and molecules.
Properties	:	They are readily absorbed by most materials and appear as heat, are reflected, refracted and diffracted.
Uses	:	In physical therapy, infrared photo-graphy and vibrational spectroscopy.

(iv) *Visible Light:*

Wavelength range	:	$\approx 7.8\times10^{-7}$ to 4×10^{-7} metre.
Production	:	By incandescent bodies, ionised gases (rearrangement of electrons in atoms and molecules) and lasers.
Properties	:	Show reflection, refraction, interference, diffraction, polarisation, photoelectric effect, photographic action and can be detected by human eye.
Uses	:	It is the basis of the science of optics and optical instruments.

(v) *Ultraviolet Radiation:*

Wavelength range	:	$\approx 4\times10^{-7}$ to 10^{-8} metre (4000-100Å).
Production	:	By sun, arc, spart, hot vacuum spark and isonised gases.
Production	:	Show diffraction, photoelectric effect, Photographic action, fluorescence, phosphorescence, ionisation.
Uses	:	In physical therapy and in the study of metals, stones, diamonds, etc.

(vi) X-rays

Wavelength range	:	$\approx 10^{-8}$ to 10^{-11} metre ($100\text{-}10^{-1}$ Å).
Production	:	By deceleration of high-energy electrons bombarding a heavy-metal target.
Production	:	They have all properties of ultraviolet rays, and are highly penetrating.
Use	:	In medical surgery, radio therapy, commerce and industry and in the study of crystal structure.

(vii) Gamma Rays

Wavelength range	:	$\approx 10^{-10}$ to 10^{-14} metre (1 to 10^{-4} Å).
Production	:	By deceleration of high-energy electrons certain nuclear reactions.
Production	:	They have all properties of γ -rays, but are more penetrating and cause serious damage when absorbed by living tissues.
Use	:	In scientific studies.

Ampere's Law

A current in a conductor produces a magnetic field. The quantitative relationship between the conduction current i and the magnetic field B produced is given by Ampere's law which is stated as

$$\oint \vec{B} \cdot d\vec{l} = \mu_0 i, \qquad \text{...(i)}$$

where μ_0 is the permeability constant and i is the current through *any* surface bounded by a closed linear path around which the line integral is taken.

Maxwell's Displacement Current: Maxwell postulated that it is not only the current in a conductor that produces a magnetic field, but a changing electric field in a vacuum, or in a dielectric, also produces a magnetic field. Thus a changing electric field is equivalent to a current which exists as long as the electric field is changing and produces the same magnetic effect as an ordinary conduction current. This pseudo-current is called the 'displacement current'. This conception of the displacement current is responsible for the development of the electromagnetic theory.

Modified Ampere's Law: Let us consider a circuit containing a battery *B*, a resistor *R*, a capacitor *C* and a tapping key *K*. When the key is pressed, a current flows and charge accumulates on the plates of the capacitor. The current decreases as the charge on the plates grows. When the capacitor is fully charged, the current stops. There is no actual flow of charge between the plates of the capacitor during charging.

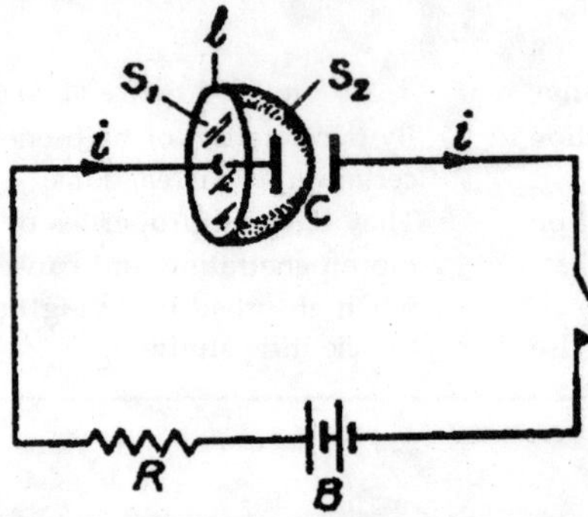

Let us consider a plane surface S_1 and a hemispherical surface S_2, both bounded by the same closed path *l*. During the process of charging the capacitor, current has been flowing through S_1 but not through S_2. Therefore, if Ampere's law is applied to the path *l* and the surface S_1 we find

$$\oint_{S1} \vec{B} \cdot d\vec{l} = \mu_0 i, \qquad \text{...(i)}$$

but if applied to the path / and the surface 5_2, then

$$\oint_{S2} \vec{B} \cdot d\vec{l} = 0.$$

The above two equations cannot both be correct. Maxwell removed this contradictory situation by adding one more term on the right-hand side of eq. (i). It is for this purpose that he regarded the changing electric field in the gap between the capacitor plates (during charging) as equivalent to a displacement current. The condition of continuity of the current requires the displacement current i_0 to be equal to $\epsilon_0 \dfrac{d\Phi_E}{dt}$

where $\in_0$ is the permittivity of the free space and Φ_E is the electric flux. Thus the Ampere's law is modified to the form

$$\oint_{S1} \vec{B} \cdot d\vec{l} = \mu_0 (i + i_d)$$

or
$$\boxed{\vec{B}.\vec{dl} = \mu_0 (i + \in_0 \frac{d\Phi_E}{dt}).} \qquad ...(ii)$$

The law is now competent to meet situations involving changing electric fields. It may be called 'Ampere-Maxwell law'.

Let us now prove that the displacement current i_d in the gap is equal to the conduction current i in the lead wires. Let q be the charge on the capacitor plates at an instant. Then the instantaneous electric field E between the plates is given by

$$E = \frac{q}{\in_0 A},$$

where A is the surface area of each plate. Differentiating it, we get

$$\frac{dE}{dt} = \frac{1}{\in_0 A}\frac{dq}{dt} = \frac{1}{\in_0 A} i,$$

where i $(=dq/dt)$ is the instantaneous current in the wires. Thus

$$i = \in_0 A \frac{dE}{dt} \qquad ...(iii)$$

Now, the displacement current *id* is, by definition,

$$i_d = \in_0 \frac{d\Phi E}{dt}$$

But $\Phi_E = EA.$

$$\therefore \quad i_d = \in_0 \frac{d}{dt}(EA) = \in_0 A \frac{dE}{dt} \qquad ...(iv)$$

Comparing cq. (iv) and (iii), we find

$$i_d = i.$$

This shows that the displacement current in the gap is equal to the real conduction current in the lead wires. Thus the conception of displacement current makes the current continuous through the entire circuit including the capacitor.

The conduction current (given by the battery) is not continuous across the capacitor gap because there is no actual flow of charge across this gap.

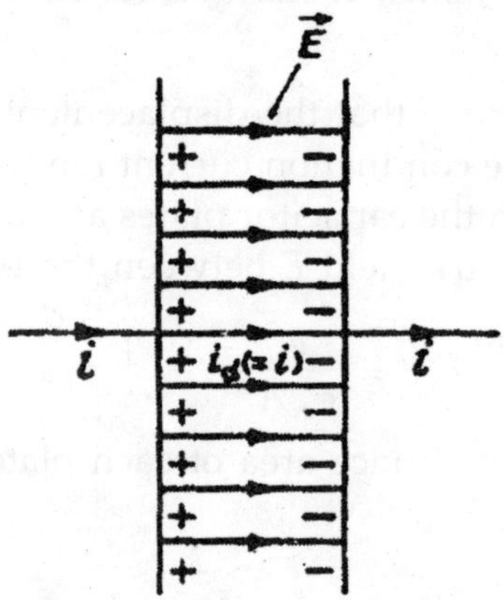

Thus the displacement current is a current only in the sense that it produces a magnetic field. It has none of the other properties of current because it is not linked with the motion of charges.

Maxwell's Equations

Maxwell, in 1862, formulated the basic laws of electricity and magnetism in the form of four fundamental equation now as 'Maxwell's equations'.

From these equations, he predicted that an accelerated charge emits transverse electromagnetic waves, consisting of fluctuating electric and magnetic fields which are mutually perpendicular to each other and travel in free space with the speed of light. He thus concluded that light itself is an electromagnetic wave.

Maxwell's equations are stated in the integral form as below :

$$\oint \vec{E} \cdot \vec{dS} = \frac{q}{\epsilon_0}. \qquad \text{...(i)}$$

$$\oint \vec{B} \cdot \vec{dS} = 0. \qquad \text{...(ii)}$$

$$\oint \vec{E} \cdot \vec{d1} = \frac{d\Phi_B}{dt}. \qquad \text{...(iii)}$$

$$\oint \vec{B} \cdot \vec{d1} = \mu_0\left(i + \epsilon_0 \frac{d\Phi E}{dt}\right). \qquad \text{...(iv)}$$

Eq. (i) is the Gauss's law for electricity which states that the electric flux through a closed surface is equal to the net charge enclosed by the surface divided by the permittivity constant ϵ_0.

Eq. (ii) is the Gauss's law for magnetism stating that the magnetic flux through a closed surface is zero. This signifies that free magnetic poles do not exist. Any volume enclosed by a surface will contain equal and opposite magnetic poles leading to a net magnetic pole-strength zero. (On the other hand, free electric charges do exist).

Eq. (iii) is the Faraday's law of electromagnetic induction, signifying that an electric field is produced by a changing magnetic flux.

Eq. (iv) is the Ampere-Maxwell law which signifies that a conduction current, as well as a changing electric flux (displacement current), produces a magnetic field.

Equation for Plane Electromagnetic Waves: In order to deduce the wave equation, we must convert the above four equations into differential form as below:

$$\frac{\partial E_x}{\partial x} + \frac{\partial E_y}{\partial y} + \frac{\partial E_z}{\partial z} = \frac{\rho}{\epsilon_0} \qquad \text{...(i)}$$

or $$\text{div } \vec{E} = \frac{\rho}{\epsilon_0}$$

$$\frac{\partial B_x}{\partial x} + \frac{\partial B_y}{\partial y} + \frac{\partial B_z}{\partial z} = 0 \qquad \text{...(ii)}$$

or $$\text{div } \vec{B} = 0$$

$$\left.\begin{aligned} \frac{\partial E_z}{\partial z} - \frac{\partial E_y}{\partial x} &= -\frac{\partial B_x}{\partial t}, \\ \frac{\partial E_x}{\partial z} - \frac{\partial E_z}{\partial x} &= -\frac{\partial B_y}{\partial t}, \\ \frac{\partial E_y}{\partial x} - \frac{\partial E_x}{\partial y} &= -\frac{\partial B_z}{\partial t}, \end{aligned}\right\} \qquad \text{...(iii)}$$

or $$\text{curl } \vec{E} = -\frac{\partial \vec{B}}{\partial t}$$

$$\left.\begin{aligned} \frac{\partial B_z}{\partial z} - \frac{\partial B_y}{\partial x} &= \mu_0\left(j_x + \epsilon_0 \frac{\partial E_x}{\partial t}\right), \\ \frac{\partial B_x}{\partial z} - \frac{\partial B_z}{\partial x} &= \mu_0\left(j_y + \epsilon_0 \frac{\partial E_y}{\partial t}\right), \\ \frac{\partial B_y}{\partial x} - \frac{\partial B_x}{\partial y} &= \mu_0\left(j_z + \epsilon_0 \frac{\partial E_z}{\partial t}\right), \end{aligned}\right\} \qquad \text{...(iv)}$$

$$\text{curl } \vec{B} = \mu_0\left(\vec{j} + \epsilon_0 \frac{\partial \vec{E}}{\partial t}\right)$$

With the help of these four equations we can relate the electric field $\vec{E}$ and the magnetic field $\vec{B}$ at any point in space to the charge density ρ and current density $\vec{j}$ respectively at that point.

For simplicity, let us derive the wave equation for vacuum, containing neither free charge nor current, *i.e.*

$$\rho = 0$$

and $$\vec{j} = 0.$$

Maxwell's equation now take the form :

$$\frac{\partial E_x}{\partial x}+\frac{\partial E_y}{\partial y}+\frac{\partial E_z}{\partial z}=0. \quad ...(i)$$

$$\frac{\partial B_x}{\partial x}+\frac{\partial B_y}{\partial y}+\frac{\partial B_z}{\partial z}=0. \quad ...(ii)$$

$$\left.\begin{aligned}\frac{\partial E_z}{\partial z}-\frac{\partial E_y}{\partial x}&=-\frac{\partial B_x}{\partial t},\\ \frac{\partial E_x}{\partial z}-\frac{\partial E_z}{\partial x}&=-\frac{\partial B_y}{\partial t},\\ \frac{\partial E_y}{\partial x}-\frac{\partial E_x}{\partial y}&=-\frac{\partial B_z}{\partial t},\end{aligned}\right\} \quad ...(iii)$$

$$\left.\begin{aligned}\frac{\partial B_z}{\partial y}-\frac{\partial B_y}{\partial z}&=\mu_0 \in_0 \frac{\partial E_x}{\partial t},\\ \frac{\partial B_x}{\partial z}-\frac{\partial B_z}{\partial x}&=\mu_0 \in_0 \frac{\partial E_y}{\partial t},\\ \frac{\partial B_y}{\partial x}-\frac{\partial B_x}{\partial y}&=\mu_0 \in_0 \frac{\partial E_z}{\partial t},\end{aligned}\right\} \quad ...(iv)$$

Let us consider the case of plane waves travelling in the x-direction, so that the wave fronts are planes parallel to the y-z plane. If the vibrations are to be represented by variations of $\vec{E}$ and $\vec{B}$, then, at any instant $\vec{E}$ and $\vec{B}$ must be constant over the whole plane. In other words, the components of $\vec{E}$ and $\vec{B}$ for this wave depend only on x and t (not on y or z). As such their partial derivatives with respect to y and z must be zero. Therefore, eq. (i), (ii), (iii) and (iv) take the following form:

$$\frac{\partial E_z}{\partial x}=0. \quad ....(1)$$

$$\frac{\partial B_x}{\partial x}=0. \quad ...(2)$$

$$\left.\begin{aligned} 0 &= -\frac{\partial B_x}{\partial t} \\ -\frac{\partial E_z}{\partial x} &= -\frac{\partial B_x}{\partial t} \\ \frac{\partial E_Y}{\partial x} &= -\frac{\partial B_z}{\partial t} \end{aligned}\right\} \qquad \ldots(3a)$$

$$\left.\begin{aligned} 0 &= \mu_0 \epsilon_0 \frac{\partial E_x}{\partial t} && \ldots(4a) \\ -\frac{\partial B_z}{\partial x} &= \mu_0 \epsilon_0 \frac{\partial E_y}{\partial t} && \ldots(4b) \\ -\frac{\partial B_y}{\partial x} &= \mu_0 \epsilon_0 \frac{\partial E_z}{\partial t} && \ldots(4c) \end{aligned}\right\}$$

Eq. (1) and (4 a) show that the longitudinal component E_x is constant in both space and time. Similarly, from eq. (2) and (3 a), B_x is also constant. These components can, therefore, have nothing to do with the wave motion, but must represent constant fields superimposed on the electromagnetic wave. For the wave, we may therefore take them to be zero, *i.e.*

$$E_x = 0, \qquad B_x = 0.$$

Consequently, $\vec{E}$ and $\vec{B}$ are both at right angles to the direction of propagation. The wave is thus transverse, and in this respect has the characteristics of light.

Of the remaining four equations, (3 c) and (4 b) involve E_y and *Bz* ; while (3 b) and (4 c) involve *E*, and B_y. For simplicity, let us confine ourselves to a polarised wave in which the electric field is along only the y-direction and the magnetic field along only the z-direction. We have then to put $E_z = B_y = 0$.

Now, we are left only with the (3*c*) and (4*b*) equations. We write them again:

$$\frac{\partial E_y}{\partial x} = -\frac{\partial B_z}{\partial t} \qquad \ldots(3c)$$

$$\frac{\partial B_z}{\partial x} = -\ \mu_0 \in_0 \frac{\partial E_y}{\partial t} \qquad ...(4c)$$

Differentiating eq. (3*c*) with respect to x and eq. (4*b*) with respect to t, we get

$$\frac{\partial^2 E_y}{\partial x^2} = -\ \frac{\partial^2 B_z}{\partial x \partial t}$$

and
$$-\frac{\partial^2 B_z}{\partial t\, \partial x} = -\ \mu_0 \in_0 \frac{\partial^2 E_y}{\partial t^2}$$

Eliminating the derivatives of B_z, we get

$$\boxed{\frac{\partial^2 E_y}{\partial x^2} = \mu_0 \in_0 \frac{\partial^2 E_y}{\partial t^2}} \qquad ...(5)$$

In a similar manner, we find

$$\boxed{\frac{\partial^2 B_Z}{\partial x^2} = \mu_0 \in_0 \frac{\partial^2 B_z}{\partial t^2}} \qquad ...(6)$$

Eq. (5) and (6) have just the form of the differential equation for plane waves $\left(\frac{\partial^2 y}{\partial x^2} = \frac{1}{v^2}\frac{\partial^2 y}{\partial t^2}\right)$ with E_y and B_z playing the part of the displacements. This means that both the electric and magnetic fields vary in time and space, as does the displacement of a Particle in a plane mechanical wave. We thus have a travelling electromagnetic wave along the *x*-direction.

Comparing eq. (5) or (6) with the differential equation of a mechanical wave, we find that the speed v_0 of the electromagnetic wave in free space is given by

$$v_0{}^2 = \frac{1}{\mu_0 \in_0}$$

$$v_0 = \frac{1}{\sqrt{(\mu_0 \in_0)}}$$

μ_0 is exactly $4\pi \times 10^{-7}$ newton/amp² by definition, and the most precisely measured value of $\in_0$ is 8.8402×10^{-12}/newton-metre². Substituting these values, we get

$$v_0 = \frac{1}{\sqrt{[(4\times 3{\cdot}1416\times 10^{-7})\times(8{\cdot}8402\times 10^{-12})]}}$$

$$= 2.9979\times 10^8 \text{m/sec},$$

which is same as the speed of light c in free space. Hence

$$\boxed{v_0 = c\,.}$$

Thus we see that eq. (3 c) and (4 b) predict the existence of a plane-polarised wave of the electric vector which lies in the x-y plane, and a plane-polarised wave of the magnetic vector which lies in the x-z plane. The two waves are interdependent (one cannot exist without the other) transverse waves, travelling together along the x-direction with the speed of light. This led Maxwell to propose that light is an electromagnetic wave.

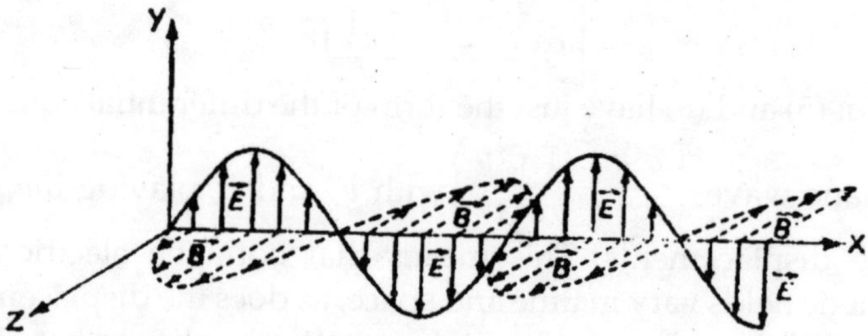

In fact, the emergence of the speed of light from purely electromagnetic considerations is the crowning achievement of Maxwell's electromagnetic theory.

The dual character of the electromagnetic wave raises the question as to whether it is the electric vector or the magnetic vector which is to be the light vector. Wiener has experimentally shown that it is the electric vector that is responsible for the photographic and fluorescent effects of light. Therefore, we can say that the electric vector is the light vector.

***Relation Between $\vec{E}$ and $\vec{B}$*:** Let us now derive relation between the magnitudes of the electric field $\vec{E}$ and the magnetic field $\vec{B}$ in an electromagnetic wave. Let us assume for simplicity that the field variations are sinusoidal. The well-known expression for a sine wave of wavelength λ, travelling in the positive x-direction with a velocity v is

$$y = a \sin\frac{2\pi}{\lambda}(vt-x).$$

On this basis, the solutions of the differential eq. (5) and (6) are of the form

$$E_y = E_m \sin\frac{2\pi}{\lambda}(ct-x)$$

and
$$B_z = B_m \sin\frac{2\pi}{\lambda}(ct-x),$$

where E_m and B_m are the maximum values (amplitudes) of $\vec{E}$ and $\vec{B}$, and $c = \dfrac{1}{\sqrt{(\mu_0 \epsilon_0)}}$

Differentiating the above two equations with respect to t and x respectively, we obtain

$$\frac{\partial E_y}{\partial t} = E_m \frac{2\pi c}{\lambda} \cos\frac{2\pi}{\lambda}(ct-x)$$

and
$$\frac{\partial E_z}{\partial x} = B_m \frac{2\pi}{\lambda} \cos\frac{2\pi}{\lambda}(ct-x).$$

Making these substitutions in eq. (4 b), we get

$$B_m \frac{2\pi}{\lambda} \cos\frac{2\pi}{\lambda}(ct-x) = \mu_0 \epsilon_0 E_m \frac{2\pi c}{\lambda} \cos\frac{2\pi}{\lambda}(ct-x)$$

or
$$B_m = \mu_0 \epsilon_0 E_m c$$

or
$$B_m = \frac{1}{c^2} E_m c \qquad \left[\because c = \frac{1}{\sqrt{(\mu_0 \epsilon_0)}}\right]$$

or
$$B_m = \frac{1}{c} E_m$$

or $$\boxed{B_z = \frac{1}{c} E_y.}$$

This is the required relatiozn, It signifies that E_y and B_z are in phase, *i.e.*, when E_y has its maximum value, B_z is also a maximum. Further, their ratio remains constant at any point.

Propagation of Energy in Electromagnetic Waves : Poynting vector

Like any other travelling wave, an electromagnetic wave transports energy from point to point. (The energy from the sun reaches the earth by electromagnetic waves). The rate of energy-flow per unit area perpendicular to the direction of propagation of a plane electromagnetic wave can be described by a propagation vector $\vec{S}$, known as Poynting vector (or Poynting flux) which is defined by

$$\vec{S} = \frac{1}{\mu_0} \vec{E} \times \vec{B}.$$

In the mks system, $\vec{S}$ is expressed in watts/metre^2, where $\vec{E}$ is in volt/metre and $\vec{B}$ in weber/metre^2. The direction of $\vec{S}$ which is perpendicular to the plane containing $\vec{E}$ and $\vec{B}$, is the direction of propagation of the wave, *i.e.*, the direction of energy-flow.

Suppose a plane electromagnetic wave is propagating in the *x*-direction. Let us consider a box of length *dx* and cross-sectional area *A* which lies in the plane of $\vec{E}$ and $\vec{B}$. The box is fixed with respect to the axes while the wave moves through it.

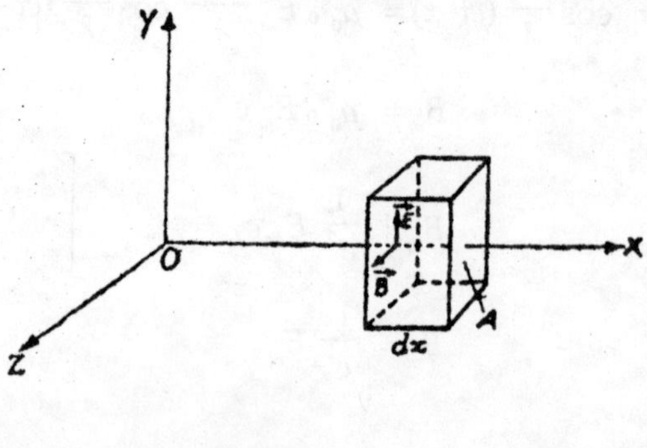

We know that the energy density in an electric field is $\frac{1}{2} \epsilon_0 E^2$, and that in a magnetic field is $\frac{1}{2\mu_0} B^2$ in the free space. Therefore, at any instant the electromagnetic energy in the box is

$$dU = \left(\frac{1}{2} \epsilon_0 E^2 + \frac{1}{2\mu_0} B^2\right) Adx, \qquad \text{...(i)}$$

where $A\,dx$ is the volume of the box and E and B are the instantaneous values of the field vectors in the box.

Now, E and B are related by

$$E = cB.$$

Replacing one of the E's in the first term of eq. (i) by cB, and *om* of the B's in the second term by E/c, we get

$$dU = \left[\frac{1}{2} \epsilon_o E(cB) = \frac{1}{2\mu_0} B(E/c)\right] Adx$$

$$= \frac{EB}{2\mu_0 c} (\mu_0 \epsilon_0 c^2 + 1) \ A \ dx.$$

But $c = \frac{1}{\sqrt{(\mu_0 \epsilon_0)}}$, so that $\mu_0 \epsilon_0 c^2 = 1$. Thus

$$dU = \frac{EB}{\mu_0 c} A \ dx.$$

Since $A\,dx$ is the volume of the box containing the energy dU, the instantaneous energy-density in the electromagnetic wave is

$$\boxed{u = \frac{dU}{Adx} = \frac{EB}{\mu_0 c}} \qquad \text{...(ii)}$$

The electromagnetic wave travels with velocity c in the x direction. If dt be the time taken by the wave to cover the length dx of the box, then $dx = c\,dt$. Thus

$$dU = \frac{EB}{\mu_0} A \ dt.$$

The energy dU now in the box will pass through the right face (of area A) of the box in a time dt. Thus the energy passing per unit area per unit time, which is the magnitude S of the Poynting vector, is given by

$$\boxed{S = \frac{dU}{Adt} = \frac{1}{\mu_0} EB.} \qquad \text{...(iii)}$$

S is also known as 'energy flux' or 'energy density. For the general case in vector notation this would be

$$\vec{S} = \frac{1}{\mu_0} \vec{E} \times \vec{B}$$

This relation gives the instantaneous rate of energy-flow per area. We can also determine the average value of S over one or more cycles of the wave. In the case of sinusoidal varying fields the time-average values of E and B are $E_m/\sqrt{2}$ and $B_m/\sqrt{2}$ where E_m and B_m are field amplitudes. Thus

$$\vec{S} = \frac{1}{2\mu_0} E_m B_m.$$

Eq. (ii) and (iii) show that the energy-density u and energy-flux a are related as

$$\boxed{u = \frac{s}{c}}$$

Momentum of an Electromagnetic Wave: Radiation Pressure

Maxwell had also predicted that electromagnetic waves transport linear momentum in the direction of propagation. It means that they exert a pressure (called radiation pressure) when falling on the surface of a body.

Suppose that a plane electromagnetic wave falls normally on a perfectly absorbing surface of area A for a time t. If energy U is absorbed during this time, the momentum p delivered to the surface is given, according to Maxwell's prediction, by

$$P = \frac{U}{c}$$

If S is the energy flux (energy passing per unit area per unit time) then $U = SAt$, so that

$$P = \frac{SAt}{c}$$

But $\frac{S}{c} = u$ (energy density).

$$\therefore \qquad p - uAt.$$

From Newton's law, the average force F on the surface is equal to the average rate at which momentum is delivered to the surface, thus

$$F = \frac{p}{t} = uA.$$

Dividing by the area A, we get the radiation pressure P_{rad} exerted on the surface.

$$p_{rad} = \frac{F}{A} = u.$$

Thus for normal incidence the radiation pressure on a perfect absorber is equal to the energy density in the wave.

On the other hand, if the surface is a perfect reflector the radiation after reflection has a momentum equal in magnitude but opposite in direction to the incident radiation. The momentum delivered will, therefore, be twice that given above, and the radiation pressure would accordingly be

$$P_{rad} = 2U.$$

3

Normal Modes of Vibration

Those coordinates are normal coordinates in which the equations of motion of a system are linear differential equations with constant coefficients, and each equation has *only one* dependent variable.

Coupled Oscillations

Degrees of Freedom: The number of independent ways in which a system can acquire energy is called the number of 'degrees of freedom' of the system. Each degree of freedom is assigned its own particular normal coordinate.

A harmonic oscillator can acquire potential energy ($\frac{1}{2}kx^2$) in terms of the normal position coordinate x and kinetic energy ($\frac{1}{2}mx^2$) in terms of the normal velocity coordinate x. Thus it has two degrees of freedom, and two normal coordinates.

Normal Modes of Vibration: A real physical system is generally capable of vibration in a number of different ways, each with a characteristic frequency. These various ways are called 'modes. A vibration involving only one dependent variable is called a 'normal mode of vibration', and the frequency of vibration in a normal mode is called 'normal frequency'. In a normal mode, all components of the system vibrate with the same normal frequency.

The normal modes of vibration are entirely independent of each other, and never exchange energy with one another. If only one mode vibrates, the other modes of the system will remain at rest.

Examples:

Two Coupled Pendulums: Let us consider a system of two identical pendulums whose bobs 1 and 2 are connected by a spring. The natural length of the spring is equal to the distance between the bobs in their equilibrium positions.

Suppose we displace the bobs 1 and 2 equally in the same direction and then release. They oscillate in phase with equal amplitudes and same frequency (which is the free natural frequency of each pendulum). The coupling spring maintains its natural length. This represents a normal mode of vibration of the system.

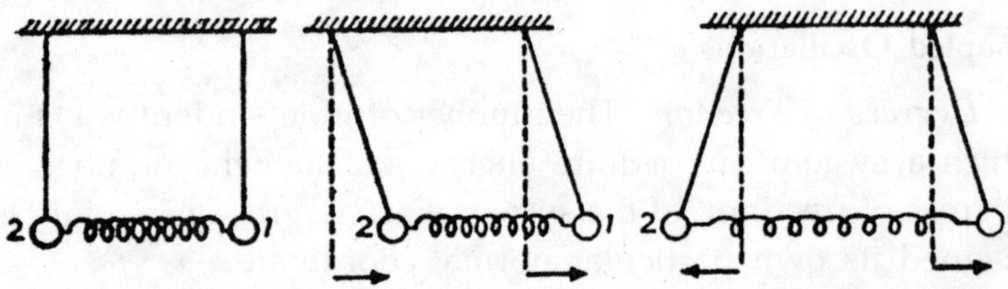

(a) (b) (c)

Let us now displace 1 and 2 equally, but in opposite directions and then release.

In this case, the coupling spring is stretched and compressed alternately, and the bobs 1 and 2 oscillate 180° out of phase with equal amplitudes and a frequency higher than that of the first mode. This is the second normal mode of vibration of the system.

Any other motion of the two coupled pendulums, in which each starts from rest, can be described as the superposition of the above two normal modes. In the general case, however, the amplitude of each pendulum varies because the two normal modes have different frequencies. The amplitude of one pendulum increases as that of the other decreases, after which the situation reverses, and so on. Thus there is an exchange of energy between the normal modes.

Triatomic Molecule

Atoms in polyatomic molecules vibrate about their equilibrium positions. Let us consider a linear triatomic molecule CO_2 having the structure O=C=O. The relative motion of the three atoms can be described in terms of normal vibrations, In figure above the two *O* atoms vibrate in phase, while the *C* atom moves in the opposite direction to conserve the position of the centre of mass.

The *O* atoms move in opposite directions, while the *C* atom remains fixed at the centre of mass. The motions are perpendicular to the line joining the atoms, resulting in a bending of the molecule.

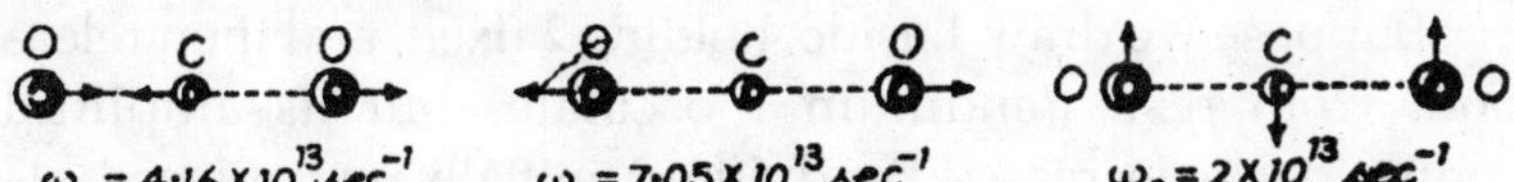

(a) (b) (c)

Each of the above normal modes has a characteristic frequency with which all the three atoms vibrate in that mode. More than one of these modes may be excited at one time and the atoms then perform complicated motions which can be analysed in normal motions.

A system consisting of two or more harmonic oscillators, interacting with one another, is called a 'coupled system' and the oscillations of such systems are called 'coupled oscillations'. During coupled oscillations, the individual oscillators exchange energy with one another because they share a common component such as stiffness (or capacitance), mass (or inductance) or resistance.

A solid body, for example, is composed of many atoms vibrating about their equilibrium positions, and the motion of each atom affects its neighbours. Thus the atoms of the solid are coupled together.

Two Coupled Pendulums: Stiffness Coupling

Let us consider a system of two identical pendulums 1 and 2, each having a mass *m* suspended on a light rod of length *l*, and connected by a light spring of force-constant (stiffness) *k*.

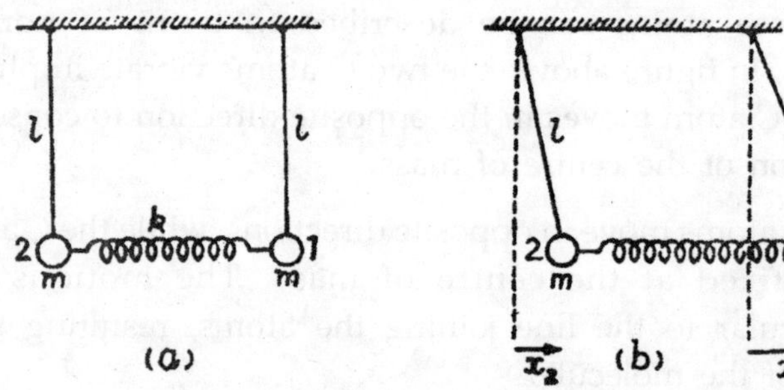

Suppose we draw 1 aside, holding 2 fixed, and then release both from rest. Pendulum 1 oscillates but its amplitude continuously decreases. Pendulum 2, initially undisplaced, also begins to oscillate with increasing amplitude. Soon 1 and 2 have equal amplitudes.

The amplitude of 1 continues to decrease and becomes zero, while that of 2 continues to increase and becomes equal to the amplitude initially given to 1. Thus the starting condition is reversed. Again the motion of 2 is transferred back to 1, and so on. The energy initially given to 1 thus shuttles back and forth between 1 and 2 through the coupling spring.

Let x_1 and x_2 be the instantaneous displacements of the Pendulums 1 and 2, and $x_1 > x_2$. The spring is stretched by an amount $x_1 - x_2$ and pulls on 1 and 2 with an elastic force $k(x_1-x_2)$. Thus the net restoring force (gravity+elasticity) on 1 is

$$-\frac{mg}{l}x_1 - k\,(x_1 - x_2)$$

and on 2 it is

$$-\frac{mg}{l}x_2 + k\,(x_1 - x_2)$$

because the elastic force (coupling term) opposes the motion of 1 but helps that of 2.

We now write the equations of motion of 1 and 2. They are

$$m\frac{d^2x_1}{dt^2} = -\frac{mg}{l}x_1 - k\,(x_1 - x_2)$$

and

$$m = \frac{d^2x_2}{dt^2} - \frac{mg}{l}x_2 - k\,(x_1 - x_2)$$

Let us write $\omega_0^2 = g/l$ where ω_0 is the natural frequency of each pendulum, and $\omega c^2 = \frac{k}{m}$ where we ωc is the coupling frequency of the spring. Then we have

$$\frac{d^2x_1}{dt^2} = \omega_0^2 - \omega_c^2(x_1 - x_2)$$

and

$$\frac{d^2x_2}{dt^2} = -\omega_0^2 + \omega_c^2(x_1 - x_2)$$

These may be written as

$$\frac{d^2x_1}{dt^2} + (\omega_0^2 + \omega_c^2)x_1 - \omega_c^2 x_2 = 0 \qquad \text{...(i)}$$

and

$$\frac{d^2x_2}{dt^2} + (\omega_0^2 + \omega_c^2)\,x_2 - \omega_c^2 x_1 = 0. \qquad \text{...(ii)}$$

Eq. (i) and (ii) are solved simultaneously. Adding them, we get

$$\frac{d^2(x_1 + x_2)}{dt^2} + \omega_0^2\,(x_1 + x_2) = 0, \qquad \text{...(iii)}$$

and subtracting eq. (ii) from eq. (i), we get

$$\frac{d^2(x_1 - x_2)}{dt^2} + (\omega_0^2 + 2\,\omega_c^2)\,(x_1 - x_2) = 0. \qquad \text{...(iv)}$$

Let us choose two new coordinates X_1 and X_2, such that

$$X_1 = x_1 + x_2$$

and $\quad X_2 = x_1 - x_2.$

Eq. (iii) and (iv) now take the form:

$$\frac{d^2x_1}{dt^2} + \omega_0^2 X_1 = 0$$

and $$\frac{d^2x_2}{dt^2} + (\omega_0^2 + \omega_c^2)\,X_2 = 0.$$

The possible solutions of these equations are

$$X_1 = C \cos d\,\omega_0{}^t \qquad \text{...(v)}$$

and $\quad X_2 = D \cos \omega' t;\quad \omega' = (\omega_0^2 + 2\,\omega_c^2)^{1/2} \qquad \text{...(vi)}$

where C and D are constants depending upon the initial conditions. (For simplicity, the initial phases are assumed to be zero).

The above equations in X_1 and X_2 represent simple harmonic oscillations of frequency ω_0 and ω' respectively. X_1 and X_2 are normal coordinates, Thus the motion of the coupled system is described in terms of two normal coordinates.

The solutions (v) and (vi). in terms of coordinates x_1 and x_2, will be

$$x_1 = \frac{1}{2}(X_1 + X_2) = \frac{1}{2} C \cos \omega_0{}^t + \frac{1}{2} D \cos \omega' t \qquad \text{...(vii)}$$

and $$x_2 = \frac{1}{2}(X_1 - X_2) = \frac{1}{2} C \cos \omega_0{}^t - \frac{1}{2} D \cos \omega' t. \qquad \text{...(viii)}$$

If the initial conditions are such that $D=0$, then $x_1 = x_2$ at all times and both pendulums would oscillate in phase with the same frequency ω_0 which is the free natural frequency of either pendulum. This is a normal mode of oscillation and ω_0 is a normal frequency of the system. In this mode the spring is always at its natural length, that is, the stiffness of the coupling has no effect.

If the initial conditions are such that C=0, then $x_1 = -x_i$ at all times and both pendulums would oscillate 180° out of phase with frequency $\omega' = (\omega_0^2 + \omega c^2)^{1/2}$ which is higher than the frequency ω_0. This is the second normal mode of oscillation and ω' is the second normal frequency of the system. In this mode, the spring is stretched and compressed alternately and the coupling is effective.

Thus the general motion of two coupled oscillations may be considered as the superposition of two normal modes of oscillation.

Hence the concept of normal modes of oscillation is of fundamental importance in the study of coupled systems.

Any configuration of the coupled system can be determined by means of the normal modes. For example, suppose the system of two pendulums is set in motion by displacing only the pendulum 1 through a distance $x_1 = A$ (say) and releasing both 1 and 2 from rest. That is, the initial conditions are

$$x_1 = A \text{ and } x_2 = 0 \text{ at } t=0.$$

Applying these conditions in eq. (vii) and (viii), we get

$$A = \frac{1}{2} C + \frac{1}{2} D$$

and $$0 = \frac{1}{2} C - \frac{1}{2} D.$$

These give: $$C = D = A.$$

Making these substitutions back in eq. (vii) and (viii), we get

$$x_1 = \frac{1}{2} A (\cos \omega_0{}^{t} + \cos \omega' t)$$

$$= A \cos\frac{(\omega' - \omega_0)'t}{2} \cos\frac{(\omega' + \omega_0)'t}{2};$$

$$x_2 = \frac{1}{2} A(\cos \omega_0{}^t - \cos \omega' t)$$

$$= A \sin\frac{(\omega' - \omega_0)t}{2} \sin\frac{(\omega' + \omega_0)t}{2}.$$

These equations show that both pendulums oscillate with the same angular frequency $\frac{(\omega' + \omega_0)t}{2}$. Their amplitudes $A \cos \frac{(\omega' - \omega_0)t}{2}$ and $A \sin\frac{(\omega' - \omega_0)t}{2}$ vary continuously in such a way that when the amplitude of one is zero, that of the other is maximum; and vice-versa. The energy initially given to 1 in displacing it shuttles back and forth between 1 and 2.

Forced Coupled Oscillator

Let us consider a system of two identical pendulums 1 and 2, each having a mass *m* suspended on a light rigid rod of length *l* and coupled by a light spring of stiffness *k*. Suppose, a harmonic driving force $F_o \cos at$ is applied to pendulum 1, while the motion of pendulum 2 is controlled only by its restoring force and the coupling spring.

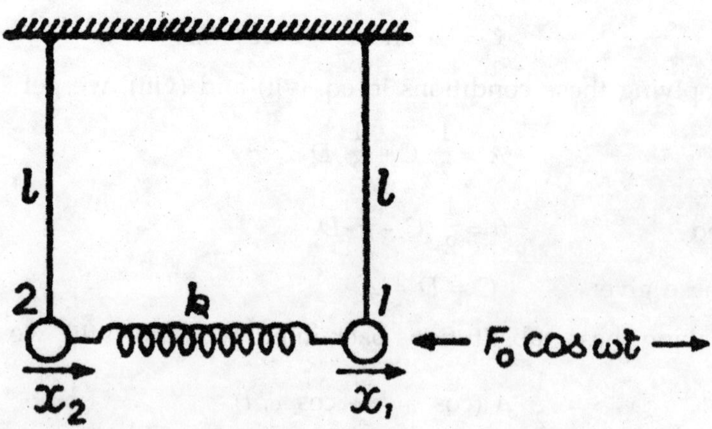

Let x_1 and x_2 be the displacements of the pendulums 1 and 2 at any instant *t*, and $x_1 > x_2$. The spring is stretched by an amount $x_1 - x_2$. The net force on 1 is

$$-\frac{mg}{l}x_1 - k(x_1 - x_2) + F_0 \cos \omega t$$

and that on 2 it is

$$-\frac{mg}{l}x_2 + k(x_1 - x_2)$$

The equations of motion of 1 and 2 are

$$m\frac{d^2x_1}{dt^2} = -\frac{mg}{l}x_1 - k(x_1 - x_2) + F_o \cos \omega t$$

$$m\frac{d^2x_2}{dt^2} = -\frac{mg}{l}x_2 + k(x_1 - x_2).$$

Writing $\frac{g}{l} = \omega_0^2, \frac{k}{m} = \omega_c^2$, we have

$$\frac{d^2x_1}{dt^2} + (\omega_0^2 + \omega_c^2)x_1 - \omega_c^2 x_2 = \frac{F_0}{m}\cos \omega t \quad \text{...(i)}$$

and $$\frac{d^2x_2}{dt^2} + (\omega_0^2 + \omega_c^2)x_2 - \omega_c^2 x_1 = 0.$$

Adding eq. (i) and (ii), we get

$$\frac{d^2}{dt^2}(x_1 + x_i) + \omega_0^2(x_1 + x_2) = \frac{F_0}{m}\cos \omega t \quad \text{...(ii)}$$

and subtracting eq. (ii) from eq. (i), we have

$$\frac{d^2}{dt^2}(x_1 - x_2) + (\omega_0^2 + 2\omega_0^2)(x_1 - x_2) = \frac{F_0}{m}\cos \omega t. \quad \text{...(iii)}$$

Introducing normal coordinates $X_1 (= x_1 + x_2)$ and $X_2 (= x_1 - x_2)$ in eq. (iii) and (iv), we have

$$\frac{d^2X_1}{dt^2} + \omega_0^2 X_1 = \frac{F_0}{m}\cos \omega t$$

$$\frac{d^2x_2}{dt^2} + \omega'^2 X_2 = \frac{F_0}{m}\cos \omega t,$$

where $\omega'^2 = \omega_0^2 + 2\omega_c{}^2$. These are the equations of undamped forced oscillations of natural frequencies ω_0 and ω'. The frequencies ω_0 and ω' are the normal mode frequencies of the system.

The steady-state solutions of the above equations are

$$X_1 = C \cos \omega t$$

and

$$X_2 = D \cos \omega t,$$

where $C = \dfrac{F_0/m}{\omega_0^2 - \omega^2}$ and $D = \dfrac{F_0/m}{\omega'^2 - \omega^2}$. The amplitudes C and D show resonances just like a single oscillator. In terms of the original coordinates, the solutions are

$$x_1 = \frac{1}{2}(X_1 + X_2) = \frac{1}{2}(C + D) \cos \omega t,$$

and

$$x_2 = \frac{1}{2}(X_1 + X_2) = \frac{1}{2}(C - D) \cos \omega t.$$

The individual amplitudes of the pendulums 1, 2, are thus given by

$$A_1 = \frac{1}{2}(C + D) = \frac{F_0}{m} \frac{\frac{1}{2}(\omega_0^2 + \omega'^2) - \omega^2}{(\omega_0^2 - \omega^2)(\omega'^2 + \omega^2)}$$

But $\omega'^2 = \omega_0^2 + 2\omega_c^2$

$$\therefore A_1 = \frac{F_0}{m} \frac{(\omega_0^2 + \omega'^2) - \omega^2}{(\omega_0^2 - \omega^2)(\omega_c{}^2 - \omega^2)} \quad ...(iv)$$

Similarly,

$$A_2 = \frac{1}{2}(C - D) = \frac{F_0}{m} \frac{\frac{1}{2}(\omega'^2_0 + \omega_0{}^2) - \omega^2}{(\omega_0^2 - \omega^2)(\omega'^2 + \omega^2)}$$

again $\omega'^2 = \omega_0^2 + 2\omega_c{}^2$

$$\therefore A_2 = \frac{F_0}{m} \frac{\omega c^2}{(\omega_0^2 - \omega^2)(\omega'^2 - \omega^2)} \quad ...(v)$$

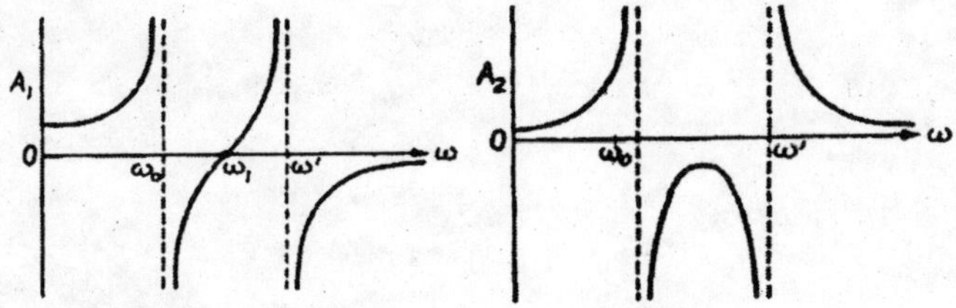

(a) (b)

The variations of A_1 and A_2, the (forced) amplitudes of the two coupled pendulums, with driving frequency ω. For frequencies upto lower resonance($\omega = \omega_0$), A_1 and A_2 are always of the same sign, *i.e.* the pendulums oscillate in phase. For frequencies beyond the higher resonance ($\omega = \omega'$), A_1 and A_2 are of opposite signs, *i.e.* the pendulums oscillate 180° out of phase. At a certain frequency $\omega_1 \left\{= (\omega_0^2 + \omega_c^2)^{1/2}\right\}$ between the resonances, A_1=0 and A_2 is non-zero. ω_1 is precisely the natural frequency of a single pendulum with coupling spring attached and the other pendulum held fixed.

4

Ultrasonics

Audible, Infrasonic and Ultrasonic Waves

The mechanical longitudinal waves can be generated in all the three forms of matter, namely, solid, liquid and gas. These waves can be divided roughly into the following three classes according to their frequency:

(i) *Audible Waves or Sound Waves:* Those longitudinal waves whose frequency lies in the range from about 20 to 20,000 Hz (sec^{1}) produce the sensation of sound on entering our ear. Hence they are called as 'audible waves' or 'sound waves'.

Audible waves are produced by vibrating bodies such as vocal cords (human and animal voice), stretched strings (violin, guitar, etc.), stretched membranes (drum, loudspeaker, etc.) air columns (flute, clarinet), and also by-irregular motions of bodies like doors, tree leaves, etc. Those audible waves which are produced by regular periodic motions of definite frequencies produce a pleasing effect on the ear and are called 'musical sound waves'. On the other hand, those waves which are produced by irregular non-periodic disturbances of bodies produce an irritating effect on the ear and are called 'noise waves'.

(ii) *Infrasonic Waves:* Those longitudinal waves whose frequency lies below 20 Hz are called 'infrasonic waves'. Our eat is not sensitive to these waves. These waves are produced by large vibrating bodies such as during an earthquake.

(iii) *Ultrasonic Waves:* Those longitudinal waves whose frequency lies above 20,000 Hz are called 'ultrasonic waves'. Our ear is not sensitive to these waves also although some animals like dogs and bats show response to them.

These waves can be produced by the high-frequency vibrations of a quartz crystal under an alternating electric field (piezo-electric effect) or by the vibrations of a ferromagnetic rod under an alternating magnetic field (magnetostriction effect).

Longitudinal waves of frequency higher than 10' Hz are called 'hypersonic waves'.

Ultrasonic Waves

The longitudinal mechanical waves having frequency above the upper limit of audibility (about 20,000 sec^{-1}) are called "ultrasonic" or "supersonic" waves. Their wavelength in air is of the order of 1 cm.

Production of Ultrasonics: There are two methods :

(i) ***Piezo-electric Method:*** When a crystal (as quartz, rochelle salt or tourmaline) is subjected to an alternating potential difference, it is set into elastic vibrations. If the frequency of the electrical oscillations coincides with one of the natural frequencies of the crystal, which are of the order of 250 to 10,000 kilocycles/ sec, a large amplitude of vibration results.

The phenomenon is utilised to produce ultrasonic waves.The necessary arrangement is shown in the figure given below. A quartz slab Q is placed between two metal plates which are connected to the anode A and the grid G of a triode valve.

Also connected to the anode A is an L-C (Inductance-

Capacitance) circuit and a hightension (H.T.) battery shunted by a bypass capacitor C_1 (which prevents high-frequency currents from passing through the battery). The variable capacitor C in the anode-circuit is then adjusted to produce electrical oscillations of frequency equal to that one of the modes of vibrations of the crystal.

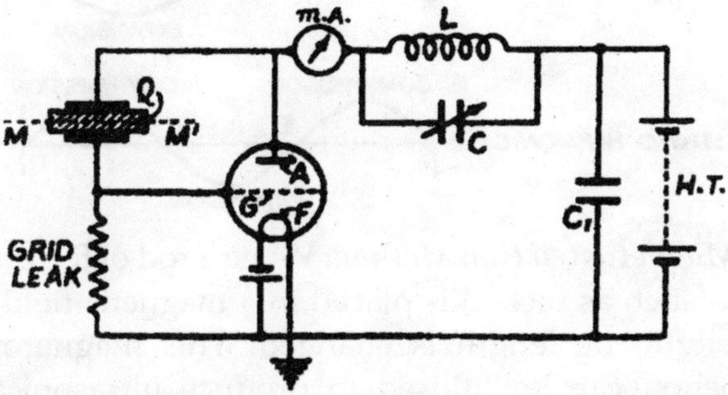

A small alternating current is set up in the anode-circuit of the valve. At some instant, the plate A is more negative and G more positive. Since A and G are connected to the opposite plates of Q, these plates are oppositely charged. The charges reverse in sign when A becomes more positive and G more negative. Thus an alternating potential difference is applied on the opposite faces of the crystal. Hence the crystal is thrown into resonant mechanical oscillations along MM' generating ultrasonic waves in the surrounding medium.

The ultrasonic generator delivers maximum power when it is operated at the fundamental frequency of the crystal. In order to generate higher-frequency ultrasonic waves, the L-C circuit is made to oscillate at a frequency equal to one of the harmonics of the crystal.

However, only *odd* harmonics of the crystal are used because piezoelectric effect can occur only when *opposite* charges appear on the electrodes. Figure given below shows the distribution of pressure and charge in the thickness of a crystal oscillating at three harmonics. It is seen that the effect will occur in the fundamental and third harmonic, and not in the second harmonic.

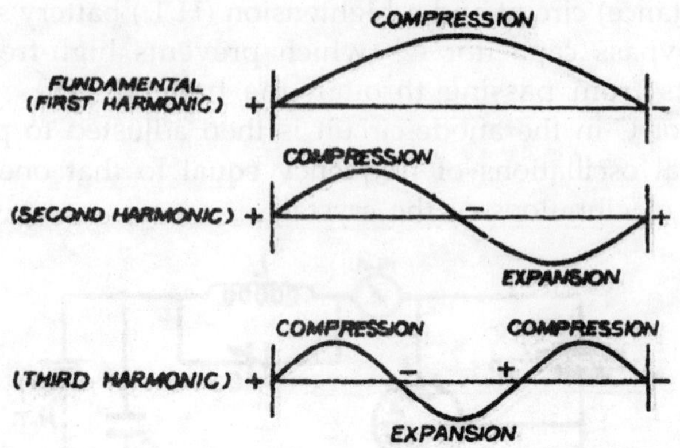

(ii) ***Magnetostriction Method:*** When a rod of ferromagnetic material (such as nickel) is placed in a magnetic field parallel to its length, its length is changed. This magnetostriction phenomenon can be utilised to produce ultrasonics. If the magnetic field is alternating and has a frequency equal to one of the natural frequencies of the rod, a large-amplitude longitudinal vibration can be set up.

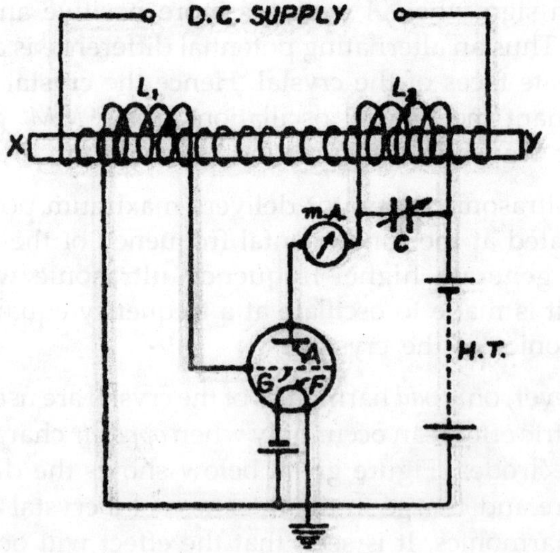

An arrangement for producing ultrasonic waves is shown

in the figure above. A short rod XY of nickel is placed in a solenoid, fed by D.C. supply, which produces a steady polarising magnetic field. Two other coils L_1 and L_2 are wound on the ends of the rod, and they are included in the grid-and anode-circuits respectively of a triode valve. The frequency of the oscillating anode-circuit is adjusted with the variable capacitor *C*. When the frequency equals the natural frequency of the rod, then the longitudinal oscillations of the rod are maintained and ultrasonics are produced in the surrounding medium.

The operation of the circuit is as follows : The rod is already magnetised by the steady polarising magnetic field. The periodic variation in the anode-current passing in the coil L_2 causes a variation in magnetisation and a consequent variation in the length of the rod. This variation in length causes a variation in the magnetic flux through the grid-coil L_1. Therefore, by the converse magnetostriction effect, an induced e.m.f. is set up in L_x. This e.m.f. acts on the grid and produces an amplified current variation in L_2. Hence the oscillations of the rod are maintained. The milli-ammetre m.A. will show an alteration when the oscillation starts. The capacitor C is adjusted to make this alteration a maximum because at this stage the frequency of the oscillating anode-current will be equal to the natural frequency of the rod and the vibration of rod will be most vigorous. A nickel rod 10 cm long, gives out ultrasonic waves of frequency 25,000 cycles per second.

Detection of Ultrasonics: We cannot directly detect the ultrasonics, although some animals, specially the bat, can do so. We can, however, detect the ultrasonics by a piezo-electric detector. When one pair of faces of a quartz crystal is subjected to ultrasonics, opposite charges are developed on the other pair perpendicular to the first. These charges are amplified by a valve amplifier and then detected by suitable means.

Properties of Ultrasonics

(i) They have a large energy content.

(ii) Their speed of propagation increases with increasing frequency.

(iii) Because of the small wavelength, the ultrasonic waves show negligible diffraction and can be transmitted over long distances without appreciable loss of energy. For this reason ultrasonic waves have been used in determining the depth of ocean by echo-sounding.

(iv) When a plane stationary ultrasonic wave is set up in a liquid, a structure is produced in which the density of the liquid varies from layer to layer along the direction of propagation of the waves. This structure can light in the same way as the structure of a crystal diffracts X-rays.

(v) Intense ultrasonic radiation has a disruptive effect on liquids by causing bubbles to be formed.

Applications of Ultrasonics

The field of the applications of ultrasonics is extremely wide. Some of the important applications are the following :

Investigation of Structure of Matter: The frequencies of ultrasonics are comparable with the frequencies of vibration and those of rotation of the molecules of matter. Hence they can be used to investigate molecular energies. The measurement of the velocity of ultrasonics in liquids and gases and its variation with frequency, temperature, etc. inform about the properties of the medium such as its compressibility, absorption, concentration, specific heat, chemical structure, arrangement of atoms in them, etc.

Detection of Flaws in Metals: The tine internal cracks and flaws in metals act as good sound reflectors of ultrasonics whose wavelength is small compared with the size of the crack or the flaw. Hence they can be detected and located by echo-sounding techniques. A part of the surface of the metal is polished and a small ultrasonic generator is placed upon it which sends a beam into the specimen. The beam will be reflected at the far surface and also at the crack or flaw if any. Near the generator is placed a receiver also which picks up the reflected echoes, which are amplified and displayed along the time-base of a Cathode-Ray-Oscillograph.

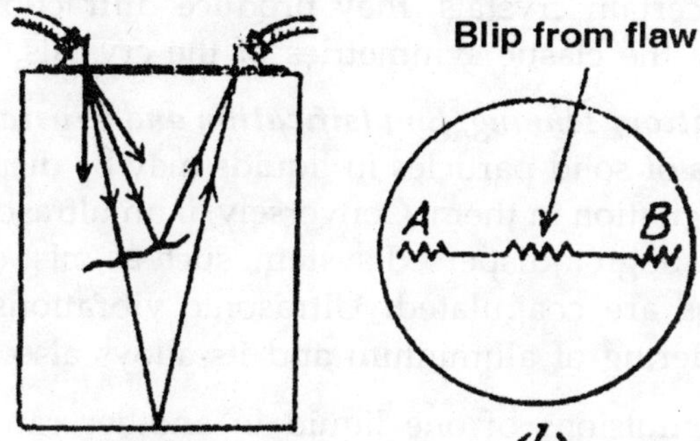

(a) (b)

The time-base shows a blip *A* in the beginning corresponding to the signal received directly, and a blip *B* at the end due to the signal reflected from the far surface. Any signal due to the flaw appears between *A* and *B*. The interval between the first and the last blip gives a length scale by which the exact location of the flaw can be known.

Directional Signalling: As the wavelength of ultrasonics is very small, it is possible to produce a short beam of ultrasonics without diffraction. Hence ultrasonic waves are used for directional signalling.

Echo Depth Sounding: Ultrasonic waves are used for determining the depth of sea, lakes, etc. by recording the time between fending an ultrasonic signal downward from the surface and receiving the echo.

Detection of Submarines, Icebergs and Other Objects in Ocean: Ultrasonics can be used to detect the reflecting objects like a submerged submarine and a hidden iceberg. The absence of diffraction effects in ultrasonic waves means that a sharp ultrasonic beam can be directed on to a small area so that detailed survey can be carried out. In this way deep and narrow valleys m the ocean floor can be detected. The location of a relatively small object on the sea-bed is also made much easier by the use of ultrasonic waves.

Elastic Symmetries of Crystals: If ultrasonic waves are applied to certain crystals, they produce diffraction images which show the clastic symmetries of the crystals.

Coagulation, Tinning, Emulsification and Crystallisation: Suspensions of solid particles in liquids may be dispersed by ultrasonic agitation in them. Conversely, if an ultrasonic beam is passed through a dispersed system, such as mist or smoke, the Particles are coagulated. Ultrasonic vibrations help in tinning soldering of aluminium and its alloys also.

Stable emulsions of one liquid in another can often be made passing ultrasonic waves into the mixture when ordinary mechanical mixing is not successful. This is used in the manufacture of the photographic films.

When a metal or alloy is crystallising from the molten state, the size of its crystals can be made smaller and more uniform by ultrasonic treatment.

Chemical Effects: In alloying certain metals, as lead and iron, it is difficult to obtain a uniform melt because the components, due to their widely different densities, tend to separate out by gravity. A beam of ultrasonics, however, keeps them well-mixed. Ultrasonics are thus useful in getting alloys of uniform composition.

Certain chemical actions are accelerated by ultrasonic waves. For example, iodine and sulphur are liberated from a solution of potassium iodide and hydrogen sulphide respectively.

Industrial Uses: Ultrasonic waves have been employed for cleaning clothes and objects like parts of watches, etc. An ultrasonic beam shakes the dirt out of the cloth into the water. In textile industry it is used to improve the dyeing properties of dye suspensions.

Biological and Medical Uses: Ultrasonics can lame or kill smaller animals like rats, fish and frogs and also bacteria. They can destroy red blood corpuscles. Unicellular plants are damped and yeast cells lose their power of reproduction when exposed to ultrasonics.

Ultrasonic vibrations have been found to produce a soothing message action on affected joints. Doctors have actually used them to cure neuralgic and rheumatic pains. Ultrasonics have also been used to localise tumours in the brain.

Determination of Velocity of Ultrasonic Waves in Liquids: Acoustic Grating: When a plane ultrasonic wave is set up in a liquid, the pressure varies with the distance of the source periodically. The density and refractive index of the liquid medium, therefore, show a periodic variation with distance from the source along the direction of propagation of the wave. If there is stationary ultrasonic wave pattern in the liquid, the density of the liquid will be greater in nodal planes than in any other plane. If, under this situation, light is passed through the liquid at right angles to the wave, the liquid behaves as a diffraction grating. Such a grating is known as "acoustic grating". The grating element is equal to the wavelength of the ultrasonic waves. Let it be λ_L . If θ be the angle of diffraction for the nth order maximum, then

$$\lambda_L \sin\theta = n\lambda,$$

where λ is the wavelength of light used, If v be the velocity of ultrasonic waves in the liquid and v be the frequency of these waves, then

$$v = \nu\lambda_L.$$

But $$\lambda_L = \frac{n\lambda}{\sin\theta}.$$

$\therefore$ $$v = \frac{\nu n\lambda}{\sin\theta}$$

From this, v can be determined.

Experimental Arrangement: This is show in the figure given below. Light from a sodium lamp is condensed by a lens *L* on the slit *Si* of a collimator *C* which renders it parallel. The parallel beam is passed through a glass cell filled with the experimental liquid.

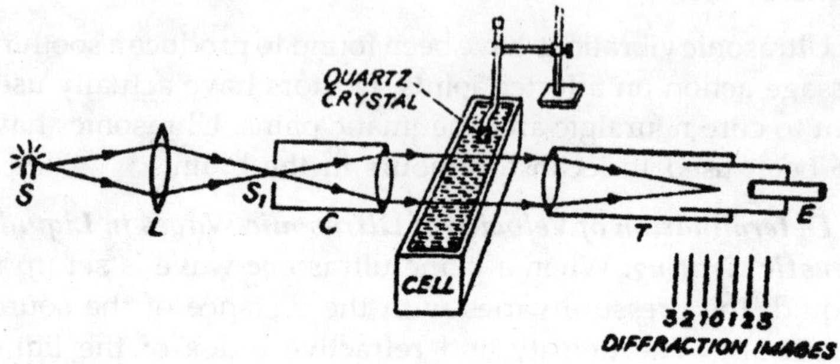

A quartz crystal fed by a radio-frequency oscillator and thus generating ultrasonic waves is suspended in the liquid such that the waves are propagated in the liquid at right angles to the light beam. The waves are reflected back from the wall of the cell and a stationary wave-pattern is set up in the liquid. The light beam passing through this pattern (acoustic grating) is diffracted. The diffracted light is received in a telescope, in the field of view of which are observed a number of diffraction images on either side of the central zero-order image.

The angular separation θ between the central (direct) image of the slit and the diffracted image of certain order *n* is measured. The frequency ν of ultrasonic waves produced by the quartz crystal is measured by means of a wave-metre. Thus, knowing θ, *n*, ν and λ (wavelength of sodium light) the velocity *v* of the ultrasonic waves can be determined by the above formula.

Bats Avoiding Obstacles

Bats have the natural power of emitting and detecting ultrasonic waves. When they fly, they continuously emit ultrasonic waves. These waves are reflected from the obstacles which happen to come in the path of the flying bats. The reflected waves are immediately detected by the bats who change their path avoiding the obstacle.

Problems

1. *Find the frequency of the first and second modes of vibration for a quartz crystal of piezoelectric oscillator. The velocity of longitudinal waves in quartz crystal is 55x10^3 m/sec.Thickness of quartz crystal is 0.05 m.*

Solution: In the lowest mode of vibration the distance between the two faces of the crystal (thickness t of crystal) will be $\lambda/2$, so that

$$\lambda = 2t = 2 \times 0.05 = 0.1 \text{ metre.}$$

Hence the lowest frequency is

$$n_1 = \frac{v}{\lambda} = \frac{5.5 \times 10^3 \text{ m/sec}}{0.1 \text{ m}} = 5.5 \times 10^4 \text{ Hz.}$$

In the second mode, the frequency would be

$$n_2 = 2n_1 = 11.0 \times 10^4 \text{ Hz.}$$

2. *A piezoelectric x-cut quartz crystal plate has a thickness of 1.6 mm. If the velocity of propagation of sound waves along the x-direction is 5760 m/s, calculate the fundamental frequency of the crystal.*

[Ans. 1.8 megahertz]

3. *A quartz crystal of thickness 0.005 m is vibrating at resonance. Calculate the fundamental frequency. Given: Y for quartz is 7.9 × 10^{10} N/m^2 and ρ for quartz is 2650 kg/m^3.*

Solution: The velocity of the longitudinal waves in the crystal is given by

$$v = \sqrt{\frac{Y}{\rho}} = \sqrt{\frac{7.9 \times 10^{10} \text{ N/m}^2}{2650 \text{ kg/m}^3}}$$
$$= 5.46 \times 10^3 \text{ m/s.}$$

The wavelength is twice the thickness of the crystal.

$$\lambda = 2t = 2 \times 0.005 \text{ m} = 0.010 \text{ m}.$$

The fundamental frequency is

$$n = \frac{v}{\lambda} = \frac{5.49 \times 10^3 \text{ m/s}}{0.010 \text{ m}} = 5.46 \times 10^5 \text{ Hz}.$$

4. *Ultrasonic pulse echo method is employed to detect possible defect in steel bar of thickness 40 cm. If the pulse arrival times are 30 and 80 microseconds, find the distance of the defect.*

Solution: Let x be the distance of the defect from the end of the bar at which the pulse enters the bar. The pulse arriving first is the one reflected from the defect and that arriving later on is the one reflected from the other end. That is, the pulse takes 30 microseconds to cover a distance of $2x$, and 80 microseconds to cover a distance of 80 cm. Hence

$$30 = \frac{2x}{v}$$

and

$$80 = \frac{80}{v}$$

Solving :

$$x = 15 \text{ cm}.$$

5. *Calculate the capacitance to produce ultrasonic waves of 10^6 Hertz with an inductance of 1 henry.*

Solution: The frequency of the *L*-*C* oscillations should be 10^6 Hertz. That is

$$n = \frac{1}{2\pi}\sqrt{\frac{1}{LC}}$$

or

$$C = \frac{1}{4\pi^2 n^2 L}$$

$$= \frac{1}{4 \times (3.14)^2 \times \left(10^6 \text{ sec}^{-1}\right)^2 \times 1 \text{ henry}}$$

$$= 0.025 \times 10^{-12} \text{ farad}$$

$$= 0.025 \ \mu\mu\text{F}.$$

5

Acoustic Impedance

Reflection and Refraction of Waves

When waves strike a boundary between two media in which they have different velocities, some of the energy is reflected back into the original medium. Also, there is a change in the direction of travel of that part of the energy which enters the new medium. This is refraction of waves. The laws of reflection and refraction can be proved by Huygen's construction of wavefront.

Reflection: Suppose a plane wavefront represented by AB, travelling in a medium with velocity v, strikes a plane boundary at A. Had there been no boundary, the wave front would have advanced to the position CD after a time t (say). But in the presence of the boundary, as the wavefront advances, the points of the boundary successively struck by the wavefront become the source of secondary spherical wavelets. Thus, after a time t, when the wavefront strikes the boundary at C, the secondary wavelet from A has acquired a radius $AE=BC=vt$, whereas the secondary wavelet from C just starts. Let us draw a tangent CE from C to the secondary wavelet from A. If CE

is a tangent common to all the secondary wavelets from the different points between A and C, then CE represents the reflected wavefront.

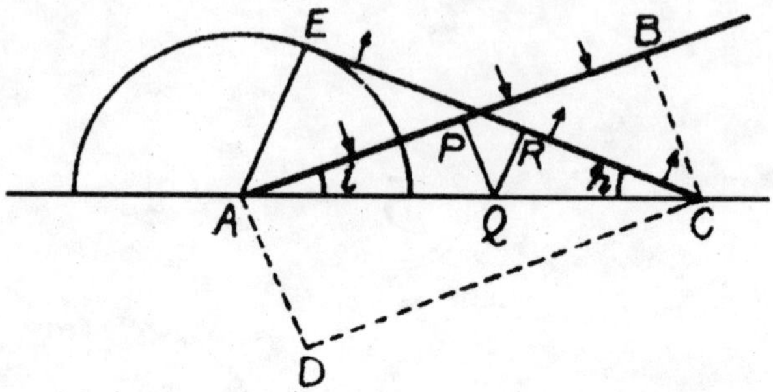

Let P be any point on AB and Q a point on AC such that PQ is perpendicular to AB. Let QR be perpendicular to CE.

Time taken for a wavelet to travel from B to $C = \frac{BC}{v}$.

Time taken for a wavelet to travel from P to R via Q

$$= \frac{PQ}{v} + \frac{QR}{v}$$

$$= \frac{AQ}{AC}\frac{BC}{v} + \frac{QC}{AC}\frac{AE}{v}$$

($\because$ in Δ^s APQ and ABC, $PQ/BC=AQ/AC$ and in Δ^s CQR and CAE, $QR/AE=QC/AC$)

$$= \frac{AQ}{AC}\frac{BC}{v} + \frac{QC}{AC}\frac{BC}{v} \qquad (\because AE = BC)$$

$$= \frac{BC}{v}\left(\frac{AQ+QC}{AC}\right)$$

$$= \frac{BC}{v}.$$

Thus the wavelets arrive at C, R and E at the same instant. Since P is any point of AB, the wavelets from all points between A and C touch CE at the same instant. Hence CE is the reflected wavefront.

The angle between the incident wavefront *AB* and the boundary *AC* is the angle of incidence *i*, and that between the reflected wavefront *CE* and the boundary *CA* is the angle of reflection *r*.

The triangles *ABC* and *CEA* are equal in all respects because $BC=EA$, $\angle ABC = \angle CEA = 90°$ and *AC* is common. Therefore,

$$\angle BAC = \angle ECA$$

or $$i = r,$$

which is the ordinary law of reflection.

Refraction: Again, let us suppose that a plane wavefront *AB* travelling in a medium with velocity v_1 strikes a plane boundary at *A*. Let its velocity in the medium on the other side of the boundary be v_2. Had there boundary, the would have advanced to the position *CD* after a time *t* (say).

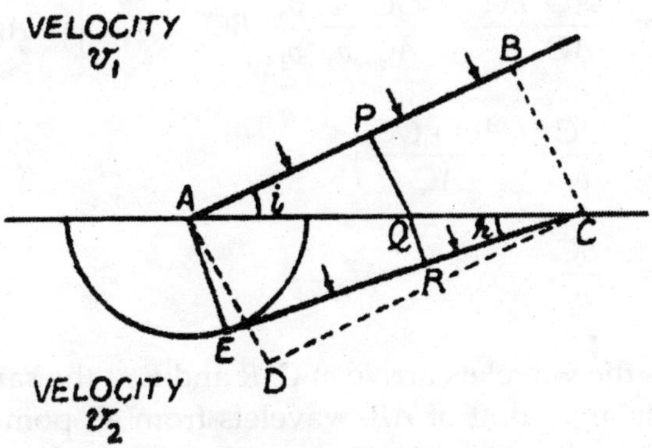

But in the presence of the boundary, as the wavefront advances, the points of the boundary successively struck by the wavefront become the source of secondary spherical wavelets. Thus, after a time *t*, when the wavefront strikes the boundary at *C*, the secondary wavelet front A has acquired a radius $AE = v_2 t = \frac{v_2}{v_1} BC$ in the new medium, where as the secondary wavelet from *C* just starts. Let us draw a tangent

CE from C to the secondary wavelet from *A* in the new medium. If *CE* is a tangent common to all the secondary wavelets from the different points between *A* and *C*, then *CE* represents the refracted wavefront.

Let *P* be any point on *AB* and *Q* a point on *AC* such that *PQ* is perpendicular to *AB*. Let *QR* be perpendicular to *CE*.

BC Time taken for a wavelet to travel from *B* to C= $\frac{BC}{v_1}$.

Time taken for a wavelet to travel from *P* .to *R* via *Q*

$$= \frac{PQ}{v_1} + \frac{QR}{v_2}$$

$$= \frac{AQ}{AC}\frac{BC}{v_1} + \frac{QC}{AC}\frac{AE}{v_2}$$

$$= \frac{AQ}{AC}\frac{BC}{v_1} + \frac{QC}{AC}\frac{1}{v_2}\frac{v_2}{v_1}BC \qquad \left[\because AE = \frac{v_2}{v_1}BC\right]$$

$$= \frac{BC}{v_1}\left(\frac{AQ+QC}{AC}\right)$$

$$= \frac{BC}{v_1}.$$

Thus the wavelets arrive at C, *R* and E at the same instant. Since *C* is any point of *AB*, wavelets from all points between *A* and *C* touch *CE* at the same instant in the new medium. Hence *CE* is the refracted wavefront.

The angle between the incident wavefront *AB* and the boundary *AC* is the angle of incidence *i*, and that between the refracted wavefront *CE* and the boundary C_A. *is* the angle of refraction *r*.

Now, $\sin i = \frac{BC}{AC}$ and $\sin r = \frac{AE}{AC}$.

$$\therefore \quad \frac{\sin i}{\sin r} = \frac{BC}{AE} = \frac{v_1}{v_2} = \text{constant.}$$

This is the ordinary law of refraction.

Characteristic Impedance of a String

When an oscillating force is applied at one end of a string, progressive waves are generated which travel along the string. The ratio of the instantaneous force at any particle of the string to the instantaneous transverse velocity of that particle is known as the "transverse impedance" or "characteristic impedance" Z of the string. Thus

$$Z = \frac{\text{transverse force}}{\text{transverse velocity}} = \frac{F}{dy/dt}.$$

This is by analogy with the electrical impedance which is the ratio of the applied alternating e.m.f. to the current produced by it. Since the driving force and the velocity are always in phase, the quantity Z is purely real (purely resistive in electrical terminology).

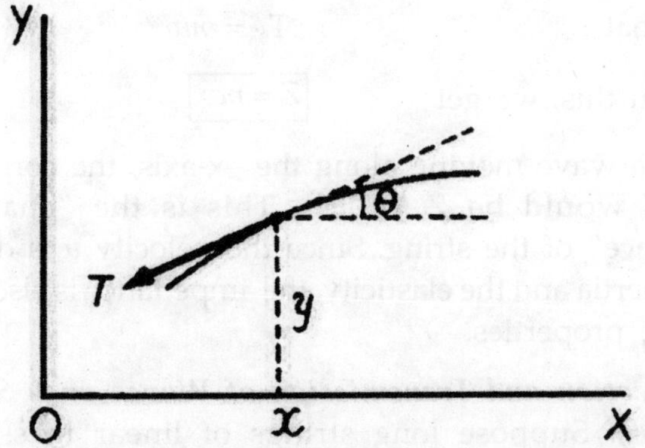

Let us consider a string stretched under a tension T along the x-axis and vibrating along the Y-axis. Suppose the tangent at a point x on the string makes an angle θ with the Y-axis at an instant when the displacements is Y. Then the transverse force acting at x would be

$$-T \sin \theta \simeq -T \tan \theta = -T\left(\frac{\partial y}{\partial x}\right),$$

assuming that θ is small, The particle velocity at x is $(\partial y/\partial t)$. Therefore, the impedance of the string is given by

$$Z = \frac{-T(\partial y/\partial x)}{\partial y/\partial t}.$$

We know the following relation between particle-velocity dy/dt and wave-velocity v, for a wave moving along the $+X$ axis:

$$\frac{\partial y}{\partial t} = -v\frac{\partial y}{\partial x}$$

This gives
$$\boxed{Z = \frac{T}{v}.}$$

Again, the wave velocity v in a string of mass per unit length m stretched under a tension T is

$$v = \sqrt{\left(\frac{T}{m}\right)}$$

so that
$$\mathrm{T} = mv^2$$

From this, we get
$$\boxed{Z = mv.}$$

For a wave moving along the –x-axis, the corresponding relation would be $Z = -mv$. This is the "characteristic impedance" of the string. Since the velocity v is determined by the inertia and the elasticity, the impedance is also governed by these properties.

Reflection and Transmission of Waves on a String at a Boundary: Suppose long strings of linear mass densities m_1 and m_2 are joined together smoothly and stretched under the same tension T along the X-axis. The wave velocities in the string will be and $v_1 = \sqrt{(T/m_1)}$ and $v_2 = \sqrt{(T/m_2)}$ their impedances will be $Z_1 = m_1 v_1$ and $Z_2 = m_2 v_2$ respectively.

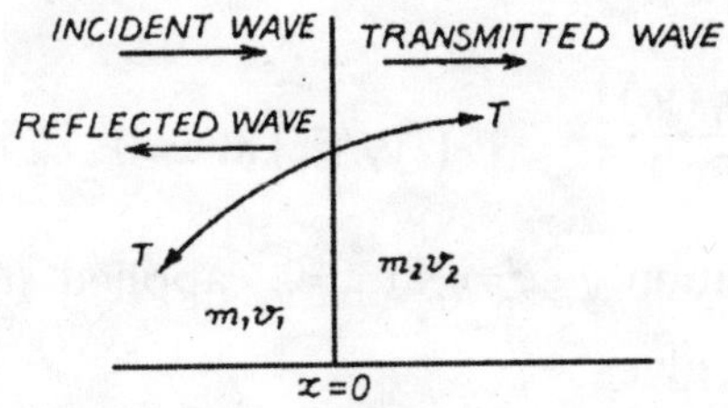

Figure

Suppose an incident plane wave travelling on the joint-string in the +X-direction reaches the boundary of two media (joint) at X= 0. A part of the incident wave is reflected from the boundary back into the medium of impedance m_1v_1 and the rest is transmitted into the second medium of impedance m_2v_2.

Let us write the equations of the incident, reflected and the transmitted waves in the following form:

$$y_i = a_i \sin \omega \left(t - \frac{x}{v_1} \right), \text{ (travelling with velocity } v_1\text{, along + X)}$$

$$y_r = a_r \sin \omega \left(t + \frac{x}{v_1} \right), \text{ (travelling with velocity } v_1\text{, along – X)}$$

and $$y_t = a_t \sin \omega \left(t - \frac{x}{v_2} \right), \text{ (travelling with velocity } v_2\text{, along +X)}$$

where a_i, a_r and a_t are the incident, reflected and transmitted amplitudes respectively. The frequency ω does *not* change with change of medium.

The conditions to be satisfied at the boundary are :

(1) The displacement at the boundary must be the same, whether we look from the first medium or from the second medium i.e., $y_i + y_r = y_t$ at $x=0$.

(2) The forces just on the two sides of the boundary must be equal, otherwise a finite difference in the forces would act on an infinitesimally small mass of the string giving an infinite acceleration. This means that

$$T\left\{\frac{\partial (y_i+y_i)}{\partial x}\right\} = T\ (\partial y_t/\partial x) \text{ at } x=0.$$

The condition $y_i+y_r=y_t$ at $x=0$. applied to the above equations, gives

$$a_i \sin \omega t + a_r \sin \omega t = a_t \sin \omega t$$

or $$a_i + a_r = a_t. \qquad \text{...(i)}$$

Now, let us differentiate the above three equations partially with respect to x:

$$\frac{\partial y_i}{\partial x} = -\frac{a_i}{v_1} \cos \omega \left(t - \frac{x}{v_1}\right),$$

$$\frac{\partial y_r}{\partial x} = -\frac{a_r}{v_1} \cos \omega \left(t - \frac{x}{v_1}\right),$$

$$\frac{\partial y_t}{\partial x} = -\frac{a_t}{v_1} \cos \omega \left(t - \frac{x}{v_2}\right)$$

The condition $T \frac{\partial}{\partial x} (y_i+y_r) = T \frac{\partial y_t}{\partial x}$ at $x=0$ gives

$$-T\frac{a_i}{v_1} \cos \omega t + T\frac{a_r}{v_1} \cos \omega t = -T\frac{a_t}{v_1} \cos \omega t$$

or $$T\frac{a_i - a_r}{v_1} = T\frac{a_t}{v_2}.$$

But $\frac{T}{v_1} = m_1 v_1 = Z_1$ and $\frac{T}{v_2} = m_2 v_2 = Z_2$, where Z_1 and Z_2 are the impedances of the two parts of the siring. Therefore

$$Z_1 (a_i - a_r) = Z_2 a_t$$

or $$a_i - a_r = \frac{Z_2}{Z_1} a_t. \qquad \text{...(ii)}$$

Adding eq. (i) and (ii), we get

$$2a_i = \left(1+\frac{Z_2}{Z_1}\right)a_t .$$

or
$$\frac{a_i}{a_t} = \frac{2}{1+\frac{Z_2}{Z_1}} = \frac{2Z_1}{Z_1+Z_2} \qquad \text{...(iii)}$$

Now, multiplying eq. (ii) by $\frac{Z_1}{Z_2}$ and subtracting from eq. (i), we get

$$(a_t + a_r) \frac{Z_1}{Z_2} (a_t - a_r) = 0$$

or
$$a_i = \left(1-\frac{Z_1}{Z_2}\right) + a_r \left(1+\frac{Z_1}{Z_2}\right) = 0$$

or
$$\frac{a_r}{a_i} = -\frac{1-\frac{Z_1}{Z_2}}{1+\frac{Z_1}{Z_2}} = \frac{Z_1-Z_2}{Z_1+Z_2}. \qquad \text{...(iv)}$$

a_r/a_i is the "reflection coefficient of amplitude" *Rc*, and a_t/a_i is the "transmission coefficient of amplitude" *Tc*. Thus, from eq. (iv) and (iii), we get

$$Rc = \frac{Z_1-Z_2}{Z_1+Z_2}$$

$$Tc = \frac{2Z_1}{Z_1+Z_2}$$

These coefficients are independent of to and hold for waves of all frequencies.

Zero Reflection: If $Z_1=Z_2$, then R_c $(= a_r/a_t)= 0$ and Tc $(= a_t/a_i) = 1$. This means that there is no reflected wave $(a_r = 0)$ and the transmitted wave has the same amplitude as the incident wave $(a_t = a_i)$.

Dependence of the Phases of Reflected and Transmitted Waves on Z_1 and Z_2: If $Z_2 > Z_1$ then Rc $(=a_r/a_i)$ is negative. This means that the reflected amplitude is opposite in sign than the incident amplitude. In other words, the phase of the displacement is changed by *it* on reflection. Tc is, however, always positive. So the transmitted wave never suffers a phase change.

Since $Rc < 1$, the reflected amplitude a_r is always smaller than the incident amplitude ar. The transmitted amplitude a_t may, however, be larger also than the incident amplitude a_t if $Z_2 < Z_1$. Of course, the energy of the transmitted wave can never be larger than that of the incident wave.

Rigid Free Boundaries are Limiting Cases: If $Z_2 = \infty$, then

$$Rc = \frac{\frac{Z_1}{Z_2} - 1}{\frac{Z_1}{Z_2} + 1} = -1$$

$$Tc = \frac{2Z_1}{Z_1 + Z_2} = 0.$$

This means that the incident wave is completely reflected with a phase change of π. This is the property of a rigid boundary (fixed end of a string or closed end of a pipe).

If $Z_2 = 0$ (no second medium), then we have

$$Rc \left(= \frac{a_r}{a_i} \right) = 1$$

and

$$\Gamma c \left(= \frac{a_t}{a_i} \right) = 2.$$

This explains the "flick" at the end of a whip when a wave reaches it.

Reflection and a Transmission of Wave Energy

Let us consider two strings of linear mass densities m_1 and

m_2. joined at a point. Let the joint string be stretched under a constant tension T along the X-axis with the joint at x=0.

Suppose a plane harmonic wave of amplitude a_i and angular frequency w is travelling along the first string at a velocity We know that the rate of flow of energy

$$= \tfrac{1}{2}\omega^2 a_i^2\, m_1\, v_1 = \tfrac{1}{2}\omega^2 a_i^2\, Z_1,$$

where Z_1 $(=m_1 v_1)$ is the impedance of the first string. This is the rate at which the energy arrives at the boundary at $x=0$.

At the boundary, the energy is partly reflected and party transmitted. Let a_r and a_t be the reflected and transmitted amplitudes respectively. If Z_2 $(=m_2 v_2)$ be the impedance of the second string, then the rate at which the energy leaves the boundary

$$= \tfrac{1}{2}\omega^2 a_r^2\, Z_1 + \omega^2 a_t^2\, Z_2$$

$$= \tfrac{1}{2}\omega^2 a_i^2 \left(\frac{a_r}{a_i}\right)^2 Z_1 + \tfrac{1}{2}\omega^2 a_i^2 \left(\frac{a_t}{a_i}\right)^2 Z_2$$

$$= \tfrac{1}{2}\omega^2 a_i^2 \left[\left\{\frac{Z_1 - Z_2}{Z_1 + Z_2}\right\}^2 Z_1 + \left\{\frac{2Z_1}{Z_1 + Z_2}\right\}^2 Z_2\right]$$

$$= \tfrac{1}{2}\omega^2 a_i^2\, Z_1 \left[\left\{\frac{Z_1 - Z_2}{Z_1 + Z_2}\right\}^2 + \left\{\frac{2Z_1}{Z_1 + Z_2}\right\}^2 \frac{Z_2}{Z_1}\right]$$

$$= \tfrac{1}{2}\omega^2 a_i^2\, Z_1 \left[\frac{(Z_1 - Z_2)^2}{(Z_1 + Z_2)^2} + \frac{4Z_1 Z_2}{(Z_1 + Z_2)^2}\right]$$

$$= \tfrac{1}{2}\omega^2 a_i^2\, Z_1 \left[\frac{(Z_1 - Z_2)^2}{(Z_1 + Z_2)^2}\right]$$

$$= \tfrac{1}{2}\omega^2 a_i^2\, Z_1$$

which is the same as the rate at which energy arrives at the boundary. Thus energy is conserved, all energy arriving at the boundary in the incident wave leaves the boundary in the reflected and transmitted waves.

The reflection and transmission energy coefficients are given by

$$\frac{\text{reflected energy}}{\text{incident energy}} = \frac{\frac{1}{2}\omega^2 a_r{}^2 Z_1}{\frac{1}{2}\omega^2 a_i{}^2 Z_1} = \left(\frac{a_r}{a_i}\right)^2 = \left(\frac{Z_1 - Z_2}{Z_1 + Z_2}\right)^2,$$

$$\frac{\text{transmitted energy}}{\text{incident energy}} = \frac{\frac{1}{2}\omega^2 a_r{}^2 Z_2}{\frac{1}{2}\omega^2 a_i{}^2 Z_1} = \left(\frac{a_t}{a_i}\right)^2 \frac{Z_2}{Z_1} = \frac{4Z_1Z_2}{(Z_1 + Z_2)^2}.$$

If the wave is incident from the second medium, the reflection coefficient would be $\left(\frac{Z_2 - Z_1}{Z_{1+} Z_2}\right)$, which is same as the above reflection coefficient. Hence the boundary reflects the same fraction of incident energy in either direction.

Impedance Matching: If $Z_1 = Z_2$, no energy is reflected and the impedances are said to be "matched". Impedance matching is very important in transfer of energy. Long-distance cables carrying energy are matched at all joints to avoid wastage by energy reflection. The power-transfer from any generator to a load is a maximum when the load matches the generator impedance. A loudspeaker is matched to the impendance of the power output of an amplifier by choosing the correct turn-ratio on the coupling transformer.

Acoustic Impedance

When an acoustic (longitudinal) wave travels in a medium, there is a variation of pressure from particle to particle. The ratio of the instantaneous pressure-excess at any particle of the medium to the instantaneous velocity of that particle is known as the "acoustic impedance" Z of the medium. Thus

$$Z = \frac{\text{pressure-excess}}{\text{particle velovity}} = \frac{p}{\partial y/\partial t}.$$

We know that $p = -K(\partial y/\partial t)$, where K is the volume elasticity of the medium.

$$\therefore \quad Z = \frac{-k(\partial y/\partial x)}{\partial y/\partial t}.$$

Now, for a wave moving in the — X direction, the particle-velocity $\partial y/\partial t$ and the wave-velocity v are related by

$$\frac{\partial y}{\partial t} = -v\frac{\partial y}{\partial x}.$$

$$\therefore \quad \boxed{Z = \frac{K}{v}.}$$

If ρ be the density of the medium, then the wave-velocity is given by

$$v = \sqrt{\left(\frac{K}{\rho}\right)}$$

or $$K = \rho v^2.$$

from this, we get

$$\boxed{Z = \rho v^2.}$$

For a wave moving in the —X direction, the corresponding relation would be $Z = -Pv$. This is the acoustic impedance of the medium and is governed by the inertial, and elastic properties of the medium.

Reflection and Transmission of Sound Wave at a Boundary between two Media: Let us consider a plane sound wave travelling in the $+X$ direction in a medium of acoustic impedance $Z_1 = \rho_1 v_1$. Let it be incident normally on an infinite plane boundary at x=0 separating the first medium from a second medium of acoustic impedance $Z_2 = \rho_2 v_2$.

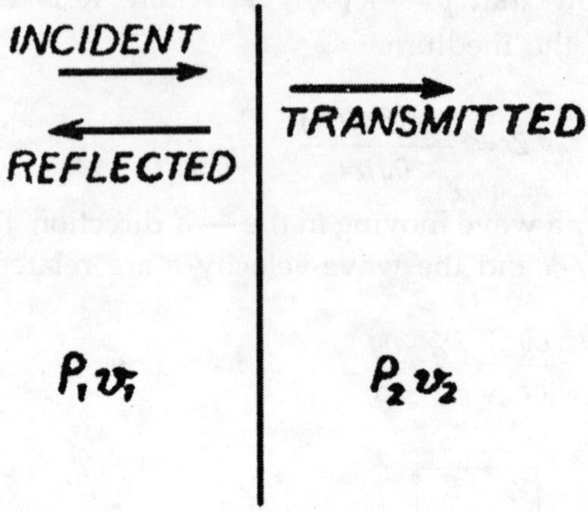

Let us write the equations of the incident, reflected and the transmitted waves in the following form :

$$y_i = a_i \sin \omega \left(t - \frac{x}{v_1}\right),$$

$$y_r = a_r \sin \omega \left(t - \frac{x}{v_1}\right),$$

and $$y_t = a_t \sin \omega \left(t - \frac{x}{v_2}\right),$$

where a_i , a_r and a_t are the incident, reflected and transmitted amplitudes respectively. The frequency ω does not change with change of medium.

The conditions to be satisfied at the boundary are :

(1) The particle velocity must be continuous across the boundary, *i.e.*

$$\left(\frac{\partial y_i}{\partial t}\right) + \left(\frac{\partial y_r}{\partial t}\right) = \left(\frac{\partial y_t}{\partial t}\right). \qquad \text{...(i)}$$

(2) The pressure-excess must also be continuous across the boundary, *i.e.*

$$p_i + p_r = p_t \cdot \qquad \text{...(i)}$$

For the incident wave

$$p_i = -K_1\left(\frac{\partial y_i}{\partial x}\right)$$

$$= -\rho_1 v_1^2\left(\frac{\partial y_i}{\partial x}\right) \qquad \left[\because v_1 = \sqrt{\frac{K_1}{\rho_1}}\right]$$

$$= \rho_1 v_1\left(\frac{\partial y_i}{\partial x}\right) \qquad \left[\because \frac{\partial y_i}{\partial t} = -v_1\frac{\partial y_i}{\partial x}\right]$$

$$= Z_1\left(\frac{\partial y_i}{\partial t}\right)$$

similarly, $p_r = -Z_1\left(\frac{\partial y_r}{\partial t}\right)$ and $p_t = Z_2\left(\frac{\partial y_t}{\partial t}\right)$. Therefore

eq. (ii) become

$$Z_1\left(\frac{\partial y_i}{\partial t}\right) - Z_1\left(\frac{\partial y_r}{\partial t}\right) = Z_2\left(\frac{\partial y_t}{\partial t}\right). \qquad \text{...(iii)}$$

Substituting the value of dy_t/dt from eq. (i) in eq. (iii), we get

$$Z_1\left(\frac{\partial y_i}{\partial t}\right) - Z_1\left(\frac{\partial y_r}{\partial t}\right) = Z_2\left\{\left(\frac{\partial y_i}{\partial t}\right) + \left(\frac{\partial y_r}{\partial t}\right)\right\}$$

or $$\frac{\partial y_i}{\partial t}(Z_1 - Z_2) = \frac{\partial y_t}{\partial t}(Z_2 + Z_1)$$

or $$\frac{\partial y_\pi/\partial t}{\partial y_i/\partial t} = \frac{Z_1 - Z_2}{Z_1 + Z_2} \qquad \text{...(iv)}$$

Substituting the value of $\partial y_r/\partial t$ from eq. (i) in eq. (iii), we get

$$Z_1\left(\frac{\partial y_i}{\partial t}\right) - Z_1\left\{\left(\frac{\partial y_t}{\partial t}\right) - \left(\frac{\partial y_i}{\partial t}\right)\right\} = Z_2\left(\frac{\partial y_t}{\partial t}\right)$$

or $$\left(\frac{\partial y_i}{\partial t}\right)(Z_1 + Z_2) = \frac{\partial y_t}{\partial t}(Z_2 + Z_1)$$

$$\frac{\partial y_t/\partial t}{\partial y_i/\partial t} = \frac{2Z_1}{Z_1 + Z_2} \qquad \text{...(v)}$$

Eq. (iv) and (v) are the expressions for the fractional reflected and transmitted particle-velocities.

Now

$$\frac{p_r}{p_i} = -\frac{Z_1(\partial y_r/\partial t)}{Z_1(\partial y_i/\partial t)} = -\frac{Z_1 - Z_2}{Z_1 + Z_2}, \qquad \text{...(iv)}$$

and $$\frac{p_t}{p_i} = -\frac{Z_2(\partial y_t/\partial t)}{Z_1(\partial y_i/\partial t)} = \frac{Z_2}{Z_1}\frac{2Z_1}{Z_1 + Z_2} = \frac{2Z_2}{Z_1 + Z_2} \qquad \text{...(iv)}$$

Eq. (vi) and (vii) are the expressions for the fractional reflected and transmitted acoustic pressures.

If $Z_1 > Z_2$, the incident and reflected particle velocities are in phase, but the incident and reflected acoustic pressures are out of phase. The reverse is the case if $Z_1 < Z_2$.

The transmitted particle-velocity and acoustic pressure are always in phase with their incident counterparts.

Reflection and Transmission of Sound Intensity: The intensity I of longitudinal waves in terms of pressure-amplitude p_m is given by

$$I = \frac{Pm^2}{2\rho v} = \frac{K^2(\partial y/\partial x)^2 m}{2\rho v} \qquad [\because p = -K(\partial y/\partial x)]$$

$$= \frac{\rho^2 v^4 (\partial y/\partial x)^2 m}{2\rho v} \qquad [\because v = \sqrt{K/\rho}]$$

$$= \frac{\rho^2 v^2 (\partial y/\partial x)^2 m}{2\rho v} \qquad \left[\because \frac{\partial y}{\partial x} = -\frac{1}{v}\frac{\partial y}{\partial t}\right]$$

$$= \tfrac{1}{2}\,\rho v\,(\partial y/\partial x)^2\, m$$

$$= \tfrac{1}{2} Z\,(\partial y/\partial x)^2\, m$$

Therefore, the intensity coefficients of reflection and transmission are respectively given by

$$\frac{I_r}{I_i} = \frac{Z_1(\partial y_r/\partial t)^2}{Z_1(\partial y_i/\partial t)^2} = \left(\frac{Z_1 - Z_2}{Z_1 + Z_2}\right)^2$$

and
$$\frac{I_t}{I_i} = \frac{Z_2(\partial y_t/\partial t)^2}{Z_1(\partial y_i/\partial t)^2} = \frac{Z_2}{Z_1}\left(\frac{2Z_2}{Z_1 + Z_2}\right)^2 = \frac{4Z_1Z_2}{(Z_1 + Z_2)^2}.$$

There is a large difference between the acoustic impedance of air and that of water. Air has a value of about 400, while water has 1.45×10^6 kg/(m^2-sec). This makes the reflection coefficient very nearly equal to unity. Hence there is an almost total (internal) reflection of sound wave energy at an air-water interface, independent of the side from which the wave approaches the interface. This is why we do not hear the noise made by sea animals under water.

6

Vibrations of Strings

Rectangular Membranes

Velocity of Transverse Waves along a Stretched String: An ideal string is an infinitely thin, uniform and perfectly flexible cord. In practice, a long and thin wire stretched under a large tension between two fixed supports is used. When the string is displaced at any point and left, transverse waves start from that point along both sides of the string. The whole string, therefore, vibrates at right angles to its len gth. The tension in the string provides the restoring force necessary for vibration. Let a string stretched under a tension T along the X-axis with one end at the origin, be set in vibration (along the Y-axis). We consider a small element AB of the string of length δx at a distance x from the origin. Let $A'B'$ be the displaced position of the element at any instant. For small displacements the tension T remains unchanged from the original tension in the stretched string.

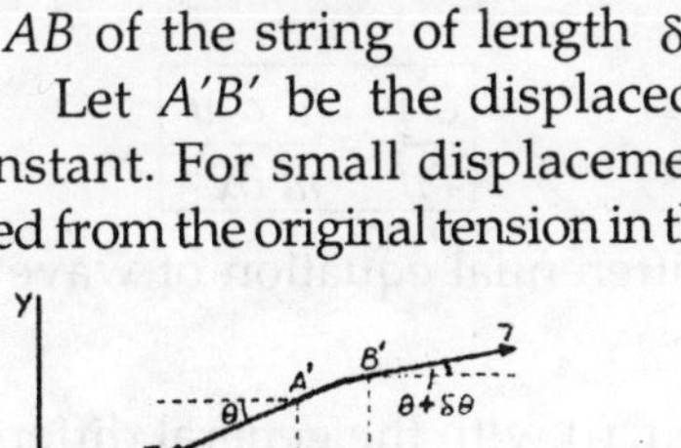

The displaced element $A'B'$ is being acted by forces (tensions) T.T tangentially at A' and B. These forces are exerted on the element by the parts of the string on the left of A' and on the right of B' respectively. Now, suppose the tangents at A' and B' make with the X-axis small angles θ *and* $\theta + \delta\theta$ respectively. Then the net force acting on the element in the X-direction is

$$T \cos (\theta + \delta\theta) - T \cos\theta,$$

which is zero, since for small angles $\cos (\theta + \delta\theta) \simeq \cos\theta \simeq 1$.

The net force on the element in the Y-direction is

$$T \sin (\theta + \delta\theta) - T \sin \theta = T \delta\theta = T\delta (\tan\theta),$$

because for small angles $\sin (\theta + \delta\theta) \simeq \theta + \delta\theta$, $\sin \theta \simeq \theta$, and $\theta = \tan \theta$. Now, $\tan \theta$ is simply the slope of the string, that is, it equals $\partial y/\partial x$. The net force in the Y-direction is thus

$$T\delta = \left(\frac{\partial y}{\partial x}\right) = T \frac{\partial}{\partial x}\left(\frac{\partial y}{\partial x}\right) \delta x = T \frac{\partial^2 y}{\partial x^2} \delta x.$$

If m be the mass per unit length of the string, then the mass of the element δx will be $m\,\delta x$. Its transverse acceleration is $\partial^2 y/\partial x^2$. By Newton's second law, the product of mass with acceleration equals the net force on the string, that is

$$(m\delta x) \frac{\partial^2 y}{\partial t^2} = T \frac{\partial^2 y}{\partial x^2} \delta x$$

or

$$\boxed{\frac{\partial^2 y}{\partial t^2} = \frac{T}{m} \frac{\partial^2 y}{\partial x^2}} \quad ...(i)$$

This is the differential equation of wave propagation in a stretched string.

Comparing eq. (i) with the general differential equation of wave motion $\frac{\partial^2 y}{\partial t^2} = v^2 \frac{\partial^2 y}{\partial x^2}$, we see that

$$v^2 = \frac{T}{m}$$

or $$v = \sqrt{\frac{T}{m}}. \qquad \text{...(ii)}$$

This is the expression for the transverse wave velocity in a stretched string.

Eq. (ii) shows that the wave velocity depends only on the tension and mass per unit length of the string, and not on the wavelength. Waves of all lengths travel with the same velocity in a given stretched string.

Equation of Motion

The equation or motion of a transverse wave in the along a string stretched in the X-direction is

$$\frac{\partial^2 y}{\partial t^2} = \frac{T}{m}\frac{\partial^2 y}{\partial x^2},$$

where T is the tension in the string and m is the mass per unit length of the string.

Normal Modes of Oscillation of a Stretched String: When a string of finite length, clamped rigidly at both ends A and B is plucked at a point P, identical waves travel along PA and PB, which are reflected at A and B. The waves travelling in opposite directions give rise to stationary waves along the string. Each point on the string oscillates transversely with constant amplitude which is a function of the position of the point. Thus the string as a whole is thrown into stationary oscillation. There are a number of well-defined states of these oscillation which are called the 'normal modes of oscillation' of the string.

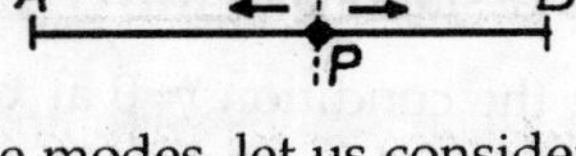

To discuss these modes, let us consider the solution of the

differential equation of the string written in the following compact form:

$$\frac{\partial^2 y}{\partial t^2} = v^2 \frac{\partial^2 y}{\partial x^2}; \qquad \text{where } v^2 = \frac{T}{m}$$

A general solution of this equation may be written as

$$y = f_1(vt-x) + f_2(vt+x).$$

y is the displacement on the string at any point x at any time t. The first part of this solution represents a wave travelling along + X–direction and the second part a wave travelling along –X-direction, both with same speed v.

Assuming waves to be simple harmonic, each of wavelength with an amplitude a in the +X-direction and an amplitude a' in the -X-direction, the above solution will be of the form

$$y = a \sin\frac{2\pi}{\lambda}(vt-x) + a' \sin\frac{2\pi}{\lambda}(vt+x), \qquad \text{...(i)}$$

with the boundary conditions that $y=0$ at $x=0$ and also at $x=l$ at all times. (This is because both ends of the string of length l are clamped.)

Applying the condition $y=0$ at x=0 for all values of t, we have

$$0 = a \sin\frac{2\pi}{\lambda} vt + a' \sin\frac{2\pi}{\lambda} vt$$

so that $a' = -a.$

Hence eq. (i) becomes

$$y = a\left[\sin\frac{2\pi}{\lambda}(vt-x) - \sin\frac{2\pi}{\lambda}(vt+x)\right]$$

$$= -2a \sin\frac{2\pi x}{\lambda} \cos\frac{2\pi x}{\lambda}. \qquad \text{...(ii)}$$

This represents a stationary wave.

Now, applying the condition $y=0$ at $x=l$ for all t, we have

$$0 = -2a \sin \frac{2\pi l}{\lambda} \cos \frac{2\pi l}{\lambda}.$$

Since this must hold for all values of t, it requires that

$$\sin \frac{2\pi l}{\lambda} = 0$$

or $$\frac{2\pi l}{\lambda} = p\pi, \qquad p=1, 2, 3, \ldots\ldots$$

or $$\lambda = \frac{2l}{p} \qquad \ldots\text{(iii)}$$

or $$n = \frac{pv}{2l} \qquad \left[\because\ n = \frac{v}{\lambda}\right]$$

where n is frequency.

But $v = \sqrt{\left(\frac{T}{m}\right)}$

$$\boxed{\therefore\ n = \frac{p}{2l}\sqrt{\left(\frac{T}{m}\right)}} \qquad P=1,2, 3, \ldots\ldots$$

This is the expression for the *permitted* frequencies of vibration of the string (obtained by putting p=1,2, 3,.......). These frequencies are 'normal frequencies' or 'normal modes' of the string. The frequency corresponding to $p=1$ is the 'fundamental' frequency of the string, while those corresponding to P=2,3 are called first, second 'overtones'.

The normal frequencies are, putting P=1, 2, 3.......

$$n_1 = \frac{1}{2l}\sqrt{\left(\frac{T}{m}\right)}$$

$$n_2 = \frac{2}{2l}\sqrt{\left(\frac{T}{m}\right)}$$

$$n_3 = \frac{3}{2l}\sqrt{\left(\frac{T}{m}\right)}$$

.. ..

... ...

We see that

$$n_1 : n_2 : n_3 : \ldots\ldots = 1 : 2 : 3 : \ldots\ldots$$

That is the frequencies form a harmonic series. Hence they are called 'harmonics' of the string. The fundamental frequency is the first harmonic, the first overtone is the second harmonic, and so on. The allowed frequencies define the length of the string as an exact number of half wavelengths. From eq. (iii), we have

$$l = p\frac{\lambda}{2} \qquad P=1, 2, 3\ldots\ldots$$

$$= \frac{\lambda}{2}, \frac{2\lambda}{2}, \frac{3\lambda}{2}, \ldots\ldots$$

Following figure shows the first three modes of vibration of the string (corresponding to $p= 1, 2, 3$). It is seen that for $p> 1$ there are $(p - 1)$ points on the string where the displacement is always zero. These are the "nodes" of the stationary waves formed along the spring.

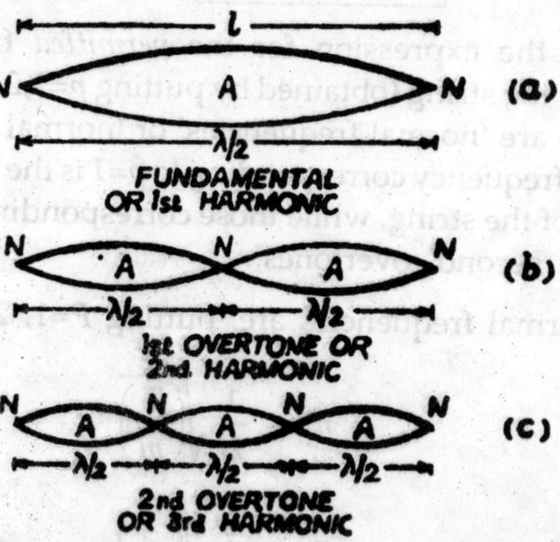

Thus the string can be made to vibrate in any of its modes by damping it (touching it lightly) at a point which is a node in that mode of vibration.

Energy of a Transverse Wave along a Stretched String

All progressive waves transmit energy through the medium in the direction of their propagation. Let us consider a transverse harmonic wave travelling along a string of mass per unit length m and stretched under a tension T along the $+x$ direction. The velocity of the wave is given by

$$v = \sqrt{\left(\frac{T}{m}\right)} \qquad \text{...(i)}$$

The equation of the wave is

$$y = a \sin \frac{2\pi}{\lambda}(vt-x), \qquad \text{...(ii)}$$

where a is the amplitude and λ is wavelength. On differentiat.ing partially with respect to t, and also twice with respect to x, we get

$$\frac{\partial y}{\partial t} = \frac{2\pi v}{\lambda} \text{ a cos } \frac{2\pi}{\lambda}(vt-x) \qquad \text{...(iii)}$$

and $$\frac{\partial^2 y}{\partial x^2} = -\frac{4\pi^2}{\lambda^2} \text{ a sin } \frac{2\pi}{\lambda}(vt-x) = -\frac{4\pi^2}{\lambda^2} y. \qquad \text{...(iv)}$$

Let us consider a small element of length δx of the string at the position x. The mass of the element is m Sx. If y is its displacement at any instant t, then its transverse velocity is $\partial y/\partial t$. Its kinetic energy at the instant t is

$$\delta \text{K} = \tfrac{1}{2}(\text{m}\,\delta x)\left(\frac{\partial y}{\partial t}\right)^2$$

Substituting the value of $(\partial y/\partial t)^2$ from eq. (iii), we get

$$\delta \text{K} = \frac{2\pi^2 m v^2}{\lambda^2} \delta x a^2 \cos^2 \frac{2\pi}{\lambda}(vt-x). \qquad \text{...(v)}$$

Suppose the string is inclined at angle θ to the x-axis at the point x at the instant t. Then the net transverse restoring force on the element δx at the instant t is

$$F = T\delta\ (\tan\theta)$$
$$= T\delta\ (\partial y/\partial x)$$
$$= T\ \frac{\partial}{\partial x}\left(\frac{\partial y}{\partial x}\right)\delta x$$
$$= -T\ \frac{4\pi^2}{\lambda^2}\, y\,\delta x \qquad \text{[from eq. (iv)]}$$
$$= -\ mv^2\ \frac{4\pi^2}{\lambda^2}\, y\,\delta x. \qquad \text{[from eq. (i)]}$$

The work done *against* the restoring force in displacing the element from 0 to y is

$$W = -\int_0^y F\, dy = mv^2\ \frac{4\pi^2 mv^2}{\lambda^2}\ \delta x \int_0^y y\, dy = \frac{2\pi^2 mv^2}{\lambda^2}\ \delta x\, y^2.$$

This is the potential energy δU of the element at the instant t. That is,

$$\delta U = \frac{2\pi^2 mv^2}{\lambda^2}\ \delta x\, y^2.$$

Substituting for y^2 from eq. (ii), we get

$$\delta U = \frac{2\pi^2 mv^2}{\lambda^2}\ \delta x\, a^2 \sin^2 \frac{2\pi}{\lambda}(vt - x). \qquad \text{...(vi)}$$

The total energy of the element is therefore, adding eq. (v) and (vi),

$$\delta E = \delta K + \delta U$$
$$= \frac{2\pi^2 mv^2}{\lambda^2}\ \delta x\, a^2.$$

So, the energy per unit length of the string is

$$E = \frac{\delta E}{\delta x} = \frac{2\pi^2 mv^2 a^2}{\lambda^2},$$

But $\frac{v}{\lambda} = n$ (frequency).

$$\therefore\ E = 2\pi^2 n^2 a^2 m.$$

It is same at all points and at all times, although kinetic and potential energies separately vary with x and t.

The waves move a distance v in one second. So all the energy contained in a length ν will flow through any fixed point in one second. Hence rate of energy flow = energy density × wave velocity

$$= 2\pi^2 n^2 a^2 \text{m}.$$

In stationary waves equal energies flow in opposite directions, Hence the net flow of energy is zero.

Laws of Vibration of Strings

When a string stretched into vibrations between two fixed points is plucked, it is thrown string vibration with nodes at the two fixed points. When the string is vibrating one loop, its natural (fundamental) frequency

$$n = \frac{1}{2l}\sqrt{\frac{T}{m}},$$

where l is the vibrating length of the string, T the tension and m per unit length. This expression suggests three laws:

(i) *Law of Length:* For a given string under a given tension the frequency varies inversely as the vibrating length of the string, *i.e.*

$$n \propto \frac{1}{l}.$$

(ii) *Law of Tension:* For a uniform string of given vibrating length and material, the frequency varies directly as the square-root of the tension, *i.e.*

$$n \propto \sqrt{T}.$$

(iii) *Law of Mass:* For a given tension and a given vibrating length of the string, the frequency varies inversely as the square-root of the mass per unit length of the string, *i.e.*

$$n \propto \frac{1}{\sqrt{m}}.$$

These laws can be verified by a sonometer.

(i) *To verify Law of Length:* A number of tuning forks of known frequencies are taken. Now a tension is applied to the sonometer wire. The length of the wire between the bridges is adjusted so that it is in unison with the note of one of the forks of frequency n_1. This length l_1 is measured. The experiment is repeated with all the forks. If l_1, l_2, l_3,...are the lengths corresponding to the forks of frequencies $n_1 n_2 n_3$...we find that

$$n_1 l_1 = n_2 l_2 = n_3 l_3 = ..$$

Alternatively, a graph between n and $\frac{1}{l}$ is plotted. This is a line straight. This shows that

$$n \propto \frac{1}{l}$$

The experiment is repeated for different constant tensions; in each case a straight line through the origin results.

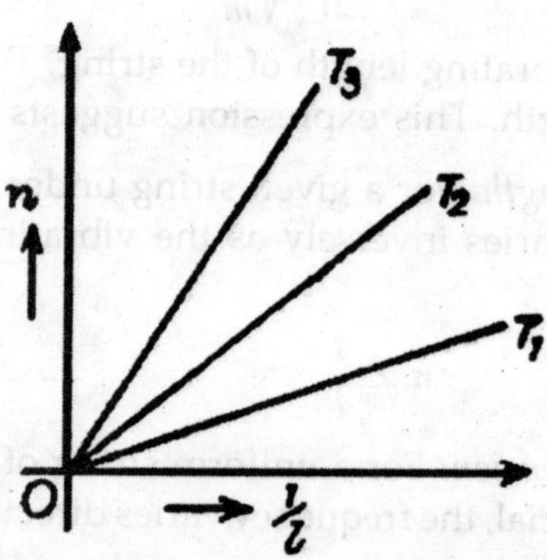

(ii) *To Verify Law of Tension:* Two wires *A* and *B* are mounted on different pairs of bridges of the same sonometer. A fixed length of the wire A between the bridges is given a small tension T_1.The other wire *B* is kept under a fixed tension. The *B* is so adjusted that its note is in unison with that of *A*. Let it be l_1. Now the tension of *A* is increased to T_2 and length of *B* adjusted until the wires, are once again in unison. Let it be l_2. This process is then repeated a number of times. Found that

$$\frac{1}{l_1} : \frac{1}{l_2} : \frac{1}{l_3} \ldots = \sqrt{T_1} : \sqrt{T_2} : \sqrt{T_3} \ldots\ldots$$

The frequencies n_1, n_2, n_3...of the wire A corresponding to its tensions. T_1, T_2, T_3...are equal to the frequencies of B corresponding to the lengths l_1, l_2, l_3... of B. Therefore, by the law of length applied to B,; $\frac{1}{l_1} : \frac{1}{l_2} : \frac{1}{l_3} \ldots = n_1 : n_2 : n_3$. Hence the last expression becomes

$$n_1 : n_2 : n_3 \ldots = \sqrt{T_1} : \sqrt{T_2} : \sqrt{T_3} \ldots$$

or $$n \propto \sqrt{T}\,.$$

If a graph between $1/l$ and $\sqrt{T}$ is plotted, a straight line through the origin is obtained, indicating that for A, $n \propto \sqrt{T}$

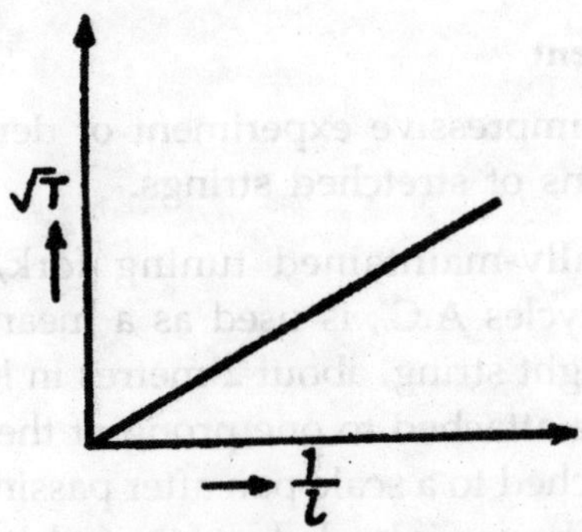

(iii) *To Verify Law of Mass:* We mount a number of experimental wires of different diameters and materials on the sonometer. The vibrating lengths of all the wires are kept equal, and they are stretched under equal tensions. Now a standard wire is mounted separately under a constant tension. The vibrating length of the standard wire is adjusted so that it is in unison with one of the experimental wires. The experiment is repeated with all the experimental wires turn by turn. Let l_1, l_2, l_3...be the lengths of the standard wire which are in unison with the experimental wires of the mass per unit length m_1, m_2, m_3........ respectively. We find that

$$l_1 : l_2 : l_3 ... = \sqrt{m_1} : \sqrt{m_2} : \sqrt{m_3} ...$$

The frequencies of the experimental wires are equal to the frequencies $n_1 : n_2 : n_3 ...$ of the standard wire corresponding to its lengths l_1, l_2, l_3 Hence by the law of length applied to the standard wire $l_1 : l_2 : l_3 ... = \frac{1}{n_1} : \frac{1}{n_2} : \frac{1}{n_3} ...$ The last expression therefore

$$\frac{1}{n_1} : \frac{1}{n_2} : \frac{1}{n_3} ... = \sqrt{m_1} : \sqrt{m_2} : \sqrt{m_3} ...$$

or
$$n \propto \sqrt{\frac{1}{m}}$$

If a graph between l and $\sqrt{m}$ is drawn, a straight line is obtained which verifies the law of mass.

Melde's Experiment

It is a very impressive experiment of demonstrating the laws of vibrations of stretched strings.

An electrically-maintained tuning fork, or a vibrator running on 50 cycles A.C., is used as a means of producing vibrations in a light string, about 2 metres in length. One end *A* of the string is attached to one prong of the fork, the other end *B* being attached to a scale pan after passing over a freely-running pulley. The pan is loaded and the fork is set in vibration.

The vibrating fork sends a continuous stream of progressive waves along the string which are successively reflected from both the ends. The length of the string between fork and pulley (or the load in the scale pan) is adjusted until the string is thrown into a stationary wave-pattern with large-amplitude antinodes. At the end *B* there is a well-defined node, while at the end *A* the amplitude is small, although the point is not quite a true node. There are two possible arrangements for performing the experiment, (i) the transverse and (ii) the longitudinal.

(i) *Transverse Arrangement:* In this arrangement the fork vibrates at right angles to the length of the string. The end of the string follows the motion of the prong step by step and this is communicated to the rest of the string. Therefore the string vibrates with the frequency of the fork.

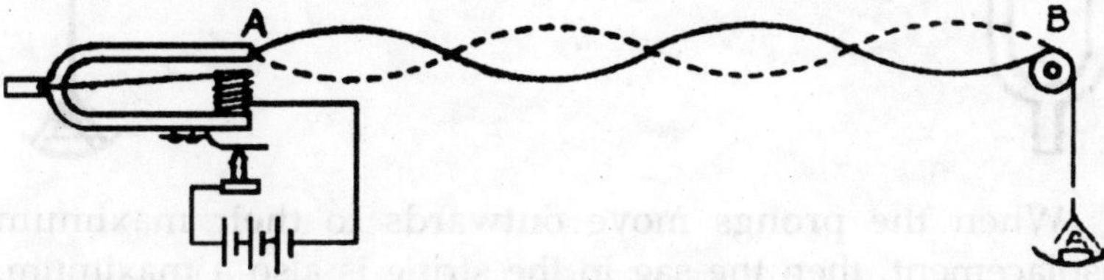

Let l be the length of the string between fork and pulley and p the number of loops in the vibrating string. Since the length of a loop is $\lambda/2$, we have

$$l = \frac{p\lambda}{2}$$

or $$\lambda = \frac{2l}{p}$$

Hence the frequency of vibration of the string is given by

$$n = \frac{p}{2l}\sqrt{\left(\frac{T}{m}\right)} \qquad \text{...(i)}$$

where T is the tension and m the mass per unit length of the string. Since n and m are fixed, the string can be made to vibrate into any number of loops by adjusting its length or tension or both When n, m and l are fixed, eq. (i) gives

$$p\sqrt{T} = \text{constant}$$

or $$Tp^2 = \text{constant.}$$

Thus for a given length of the string, if the tension is decreased to $\frac{1}{4}$ th of its previous value, the number of loops is doubled, if $\frac{1}{9}$ th tension is decreased to the its previous value, the number of loops becomes three times, and so on.

(ii) ***Longitudinal Arrangement:*** In this arrangement the fork vibrates along the length of the string. In this case the string vibrates with half the frequency of the fork as seen from the following consideration:

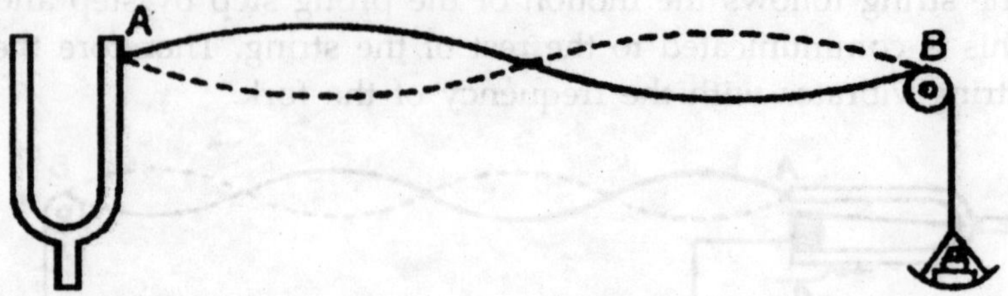

When the prongs move outwards to their maximum displacement, then the sag in the string is also a maximum. As the prongs move inwards, the string begins to move upwards. When the prongs have their maximum inward displacement, the string is pulled taut having maximum velocity in the upward direction as indicated by arrows. As the prongs again move outwards, the string continues to' move upwards under its inertia, so that when the prongs are at their maximum outward displacement, the string bows upwards. Thus for one complete oscillation of the fork the string makes half an oscillation. Hence the frequency of the string is one-half of that of the fork, *i.e.* $\frac{n}{2}$.

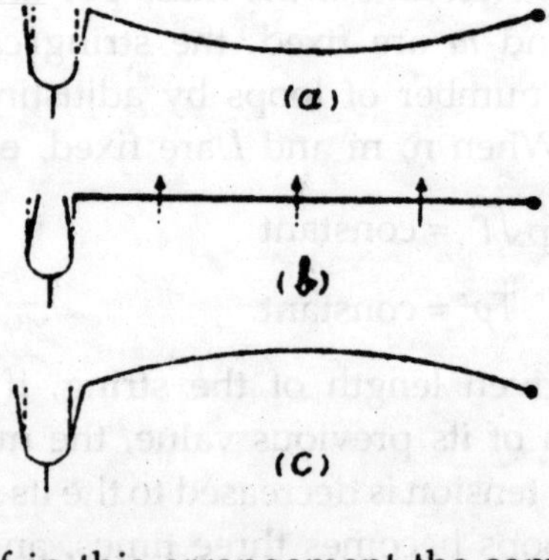

Therefore, if in this arrangement the same length l of the string is to vibrate in p loops, we have to apply tension T' such that

$$\frac{n}{2} = \frac{p}{2l}\sqrt{\left(\frac{T'}{m}\right)} \qquad ...(ii)$$

From eq. (i) and (ii), we get

$$T = \frac{T'}{4}.$$

This shows that to obtain the same number of loops in the same length of the string in the longitudinal arrangement as in the transverse arrangement, the tension must be reduced to one quarter of its value. This also means that in the longitudinal arrangement the same length of the string under the same tension will vibrate in half the number of loops than that in the transverse arrangement. All these facts indirectly verify the laws of vibration of strings.

Wave Equation in Two Dimensions

Modes of Vibration of a Membrane: A membrane is a very thin, perfectly flexible, and uniform film of material, stretched in all directions by a force which is unaffected by the motion of the membrane. In practice, membranes are used in drums and tambourines, where they have a fixed boundary along which the tension is applied uniformly.

This tension alone supplies the restoring force necessary for the transverse vibrations of the membrane. In this respect the (two-dimensional) membrane is analogous to a (one-dimensional) stretched string.

Let there be a rectangular membrane lying in the *X*-*Y* plane with the origin of coordinates at one corner of the rectangle. Let *a* and *b* be the lengths of the sides of the rectangle in the *X* and *Y* directions respectively. Let *T* be the tension *per unit length* of the boundary and *m* the mass *per unit area* of the membrane. As the membrane is set in vibration, the displacements take place at right angles to the *X*-*Y* plane, that is, in the Z-direction.

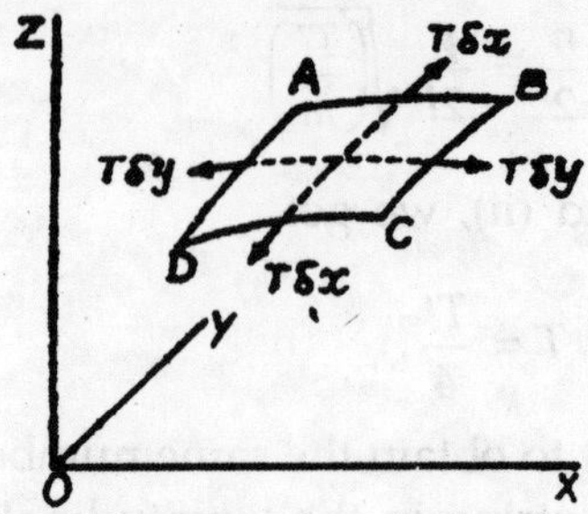

Figure given above shows a small rectangular element *ABCD* of sides *AB* and *BC* of lengths δx and δy respectively of the vibrating membrane. The forces $T\,\delta x$ and $T\,\delta y$ acting along its sides δy and δx tend to restore the element to its equilibrium position. The force $(T\,\delta x)$ acting along δy produces a force

$$(T\delta x)\,\frac{\partial^2 z}{\partial y^2}\,\partial y$$

in the z-direction; while the force $(T\delta y)$ acting along δx produces a force

$$(T\delta y)\,\frac{\partial^2 z}{\partial x^2}\,\partial x$$

also in the z-direction. The sum of these restoring forces in the z-direction is equal to the mass of the element $m\;\delta x\;\delta y$ multiplied by the acceleration of the element in the z-direction (Newton's second law). That is,

$$(T\delta x)\,\frac{\partial^2 z}{\partial y^2}\,\partial y + (T\delta y)\,\frac{\partial^2 z}{\partial x^2}\,\partial x = m\,\delta x\,\delta y\,\frac{\partial^2 z}{\partial t^2}$$

or

$$\boxed{\frac{\partial^2 z}{\partial t^2} = \frac{T}{m}\left(\frac{\partial^2 z}{\partial x^2} + \frac{\partial^2 z}{\partial y^2}\right).} \qquad \text{...(iii)}$$

This is the wave equation in two dimensions. The wave velocity is $v = \sqrt{(T/m)}$ as in case of stretched string, but m is now the surface density.

Since the boundary of the membrane is fixed, the boundary conditions are

$$z=0 \text{ at } x=0 \text{ and also at } x=a$$

and $$z=0 \text{ at } y=0 \text{ and also at } y=b.$$

The solution of the wave equation (i), taking note of the boundary conditions will be of the following form :

$$z=C \sin \frac{h\pi x}{a} \sin \frac{k\pi y}{b} \cos \omega t, \quad \text{...(ii)}$$

where h=l, 2, 3........ and k=l, 2, 3,.....

Differentiating eq. (ii) partially twice with respect to t, x and y separately, we have

$$\frac{\partial^2 z}{\partial t^2} = -C\omega^2 \sin \frac{h\pi x}{a} \sin \frac{k\pi y}{b} \cos \omega t = -\omega^2 z$$

$$\frac{\partial^2 z}{\partial x^2} = -C\frac{h^2\pi^2}{a^2} \sin \frac{h\pi x}{a} \sin \frac{k\pi y}{b} \cos \omega t, = -\frac{h^2\pi^2}{a^2} z$$

and $$\frac{\partial^2 z}{\partial y^2} = -C\frac{k^2\pi^2}{b^2} \sin \frac{h\pi x}{a} \sin \frac{k\pi y}{b} \cos \omega t, = -\frac{k^2\pi^2}{b^2} z.$$

Substituting these values in eq. (i), we get

$$-\omega^2 z = \frac{T}{m}\left(-\frac{h^2\pi^2}{a^2}z - \frac{k^2\pi^2}{b^2}z\right)$$

or $$\omega^2 = \frac{T}{m}\pi^2\left(\frac{h^2}{a^2}+\frac{k^2}{b^2}\right).$$

The corresponding frequency expression is

$$n_{h,k} = \frac{\omega}{2\pi} = \frac{1}{2}\sqrt{\left[\frac{T}{m}\left(\frac{h^2}{a^2}+\frac{k^2}{b^2}\right)\right]}. \quad \text{...(iii)}$$

This gives the possible frequencies (corresponding to various values of h and k) of the membrane. These frequencies are 'normal frequencies' or 'normal modes of vibration' of the membrane. They do not show simple numerical relationships, as we observe in the normal frequencies of a string.

The fundamental vibration is given by h=1, k=1, so that

$$n_{1,1} = \frac{1}{2}\sqrt{\left[\frac{T}{m}\left(\frac{1}{a^2}+\frac{1}{b^2}\right)\right]}.$$

This is the (lowest) frequency of the fundamental tone of the membrane. In fact, the lower overtones lie close together so that the general note of the membrane is a noise in which the fundamental tone predominates.

In the general mode (h, k) zero displacement or nodal lines are given by

$\sin\frac{h\pi x}{a} = 0$ and $\sin\frac{k\pi y}{b} = 0$.

Thus nodal lines occur at

$$x = 0, \frac{a}{h}, \frac{2a}{h}, \ldots\ldots\ldots a$$

and $$y = 0, \frac{b}{k}, \frac{2b}{k}, \ldots\ldots\ldots b.$$

Thus the nodal lines for (1,1) mode are at x=0, a and y=0, b: for (2, 1) mode at x=0, $\frac{a}{2}$, a and y=0, b ; for (3, 1) mode at x=0, $\frac{a}{3}$, $\frac{2a}{3}$, a and y=0, b ; and so on. The displacements on opposite sides of any nodal line have opposite sign. A few normal modes are shown in the following figure, in which the shaded and plain areas have opposite displacements.

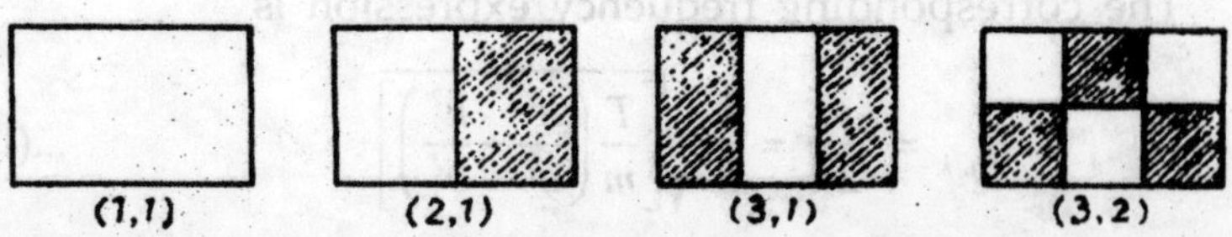

Degenerate Modes: The membrane may vibrate in different modes according to the values of h and k, with different frequencies. It is, however, possible that two or more modes may have the same frequency. Such modes are called 'degenerate modes.'

For example, the frequencies of a square membrane (a=b) are given by

$$n_{h,k} = \frac{1}{2}\sqrt{\left[\frac{T}{m}\left(\frac{h^2+k^2}{a^2}\right)\right]}$$

We can see that the modes in which the values of *h* and *k* are interchanged have the same frequency. For example, the modes (1, 2) and (2, 1) both have frequency $\frac{1}{2}\sqrt{\left(\frac{5T}{ma^2}\right)}$. This means that the same frequency corresponds to two geometrically distinct patterns of vibration. Similarly, the modes (4, 7) (7, 4) (1, 8) and (8,1) all have same frequency.

Degenerate modes are found in rectangular membrane also when the ratio of the lengths of two sides is a ratio of integers. For example, if *a=2b*, the modes (2, 2) and (4, 1) have same frequency.

Problems

1. What is the speed of transverse wave in a rope of length 1 m and mass 0.6 kg under a tension of 1000 N ?

Solution: The speed *v* of transverse waves in a string of mass *m per unit length* (linear density) stretched under a tension *T* is given by

$$v = \sqrt{\left(\frac{T}{m}\right)}.$$

Here, T=1000 *N* and $m = \frac{0.6}{1}$ =0.6 kg/m.

$$\therefore \quad v = \sqrt{\left(\frac{1000\ \text{N}}{0.6\ \text{kg/m}}\right)} = 41 \text{ m/s}.$$

2. Calculate the speed of transverse wave in a wire of radius 0.05 cm under the tension produced by 1.0 kg weight. The density of the material of wire is 9.8 gm/cm³.

Solution: The speed of transverse waves in a string of mass m per unit length under a tension T is given by

$$v = \sqrt{\left(\frac{T}{m}\right)}.$$

Here, T = 1.0 kg-wt = 1.0 × 9.8 N [$\because$ 1 kg-wt = g newton]

and m = cross-section area $\left(\pi r^2\right)$ × density

$$= 3.14 \times (0.05 \times 10^{-2}\ \text{m})^2 \times (9.8 \times 10^3\ \text{kg/m}^3)$$

$$= 7.7 \times 10^{-3}\ \text{kg/m}.$$

$$\therefore \quad v = \sqrt{\left(\frac{10 \times 9.8\ \text{N}}{7.7 \times 10^{-3}\ \text{kg/m}}\right)} = 35.7\ \text{m/s}$$

3. *Calculate the speed of transverse waves in a wire of 1.0 mm² cross-section under the tension produced by 0.1 kg-wt. The density of the material of the wire is 9.81 gm/cc. and g= 9.81m/S².*

[Ans. 10 m/s]

4. *Consider a rope 10 metres long with a mass of 0.050 kg. One end of the rope is given a transverse motion of small amplitude with a frequency of 10 sec⁻¹. If the tension in the rope is 200 N what is have wavelenght of the resultant wave?* *[Ans. 20 m]*

5. *Calculate the tensile stress in a steel wire of density 7.5 gm/cm³ when the velocity of transverse wave in the wire is 400 m/s.*

Solution: The transverse wave velocity is

$$v = \sqrt{\left(\frac{T}{m}\right)}.$$

so that the tension (force) in the wire of mass m per unit length is given by

$$T = v^2 m.$$

If A be the area of cross-section of the wire, then

$$\text{tensile stress} = \frac{T}{A} = \frac{v^2 m}{A}.$$

Now, m = $A \times \rho$, where ρ is the density of the material of the wire.

$$\therefore \quad \text{tensile stress} = \frac{v^2 A\rho}{A} = v^2 \rho$$

$$= (400{,}00 \text{ cm/sec})^2 \times (7.5 \text{ gm/cm}^3)$$

$$= 1.2 \times 10^{10} \text{ dynes/cm}^2.$$

6. A string of mass 2.0 gm/metre carries progressive waves of amplitude 1.5 cm, frequency 60 sec $^{-1}$ and speed 200 metre/sec. Calculate the energy per metre length of the wire, and the rate of energy propagation in the wire.

Solution: The energy *per metre length* of the wire is given by, in usual notations,

$$E = 2\pi^2 n^2 a^2 m.$$

Putting the given values :

$$E = 2 \times (3.14)^2 \times (60 \text{ sec}^{-1})^2 \times (1.5 \times 10^{-2} \text{ m})^2 \times (2.0 \times 10^{-3} \text{ kg/m})$$

$$= 3.2 \times 10^{-2} \text{ joule/metre.}$$

Rate of energy propagation=energy per unit length × wave speed

$$= 3.2 \times 10^{-2} \text{ joule/metre} \times 200 \text{ metre/sec}$$

$$= 6.4 \text{ watt.}$$

7. A flexible string of length 0.99 m and mass 5 gm is stretched by a tension of T newton The string vibrates in five segments with a frequency of 1000 Hz. Calculate the tension in the string.

Solution: The frequency of a wire of length l, mass per unit length m, stretched under a tension T and vibrating in p loops is given by

$$n = \frac{p}{2l}\sqrt{\left(\frac{T}{m}\right)},$$

The tension is thus given by

$$T = \frac{4l^2 n^2 m}{p^2}.$$

Putting the given values, we get

$$T = \frac{4 \times (0.99\ \text{m})^2 \times (1000\ \text{Hz})^2 \times \left(\frac{5 \times 10^{-3}}{0.99}\frac{\text{kg}}{\text{m}}\right)}{(5)^2}$$

$$= 792\ \text{N}.$$

8. A wire gives out a fundamental note of 256 cycles per sec, when it is under a tension of 10 kg-wt. Under what tension will the string emit a frequency of 512 cycles per sec?

Solution: The fundamental frequency of a stretched wire is given by

$$n = \frac{1}{2l}\sqrt{\left(\frac{T}{m}\right)}$$

where l is the length, T the tension and m the linear density of the wire. For a given wire, l and m are fixed. Thus

$$n \propto \sqrt{(T)}$$

or $$T \propto n^2.$$

Hence for doubling the frequency, the tension should be increased to 4 times its initial value, *i.e.* the new tension should be 40 kg-wt.

9. A given 90-cm long wire of sonometer is divided in three parts by two bridges whose fundamental frequencies are in the ratio 3:4:6. Find the positions of the bridges.

Solution: By the law of length of stretched strings, we have

$$n_1 l_1 = n_2 l_2 = n_3 l_3 .$$

Given that $n_1 : n_2 : n_3 = 3 : 4 : 6.$

The least common multiple is 12, so that we must have

$$l_1 : l_2 : l_3 = 4 : 3 : 2.$$

The wire is 90-cm long. Hence to divide it in three parts

in ratio 4 : 3 : 2, the bridges must be placed at 40 cm and 70 cm.

10. A wire under tension vibrates with a frequency of 450 per second. What would be the fundamental frequency if the wire were half as long, twice as thick and under one-fourth tension.

Solution: The frequency n of a wire of length l, mass per unit length m, stretched by a tension T and vibrating in p segments, is given by

$$n = \frac{p}{2l}\sqrt{\left(\frac{T}{m}\right)}$$

For emitting lowest characteristic frequency (450) it should vibrate in one segment, *i.e.* p=1. Also $m = \pi r^2 d$ where r is the radius and d the density. Therefore, we have

$$\text{n} = 450 = \frac{1}{2l}\sqrt{\left(\frac{T}{\pi r^2 d}\right)}$$

or $$450 = \frac{1}{2lr}\sqrt{\left(\frac{T}{\pi d}\right)}$$

Now, let n' be the frequency when l is reduced to $l/2$. r is increased to $2r$ and T is reduced to $T/4$. Thus

$$n' = \frac{1}{2(l/2)2r}\sqrt{\left(\frac{T/4}{\pi d}\right)}$$

$$= \left(\frac{1}{2}\right)\left(\frac{1}{2lr}\right)\sqrt{\left(\frac{T}{\pi d}\right)}$$

Comparing with eq. (i), we see that

$$\therefore \quad n' = \tfrac{1}{2} \times 450 = 225.$$

11. Four violin strings, all of the same length and material, but of diameters in the ratio 4:3:2:1 are to be stretched so that each gives a note whose frequency is § times that of the preceding string. If the stretching force of the first string is 2.048 kg-wt, calculate the tensions in the other strings.

Solution: Let T_1 , T_2, T_3, T_4 be the tensions and $4r$, 3r, $2r$, r the radii of the four strings. Their length l and density of material d are same. In view of the given data, let their frequencies be respectively n, (3/2) n, (9/4) n and (27/8) n. Then we have

$$n = \frac{1}{2l}\sqrt{\left(\frac{T_1}{\pi(4r)^2 d}\right)}, \quad \text{...(i)}$$

$$\frac{3}{2}\, n = \frac{1}{2l}\sqrt{\left(\frac{T_2}{\pi(3r)^2 d}\right)}, \quad \text{...(ii)}$$

$$\frac{9}{4}\, n = \frac{1}{2l}\sqrt{\left(\frac{T_3}{\pi(2r)^2 d}\right)}, \quad \text{...(iii)}$$

and $$\frac{27}{8}\, n = \frac{1}{2l}\sqrt{\left(\frac{T_4}{\pi(r)^2 d}\right)}, \quad \text{...(iv)}$$

Dividing eq. (ii) by (i), we have

$$\frac{3}{2} = \frac{4}{3}\sqrt{\left(\frac{T_2}{T_1}\right)}$$

or $$\frac{9}{4} = \frac{16}{9}\left(\frac{T_2}{T_1}\right)$$

$$\therefore \quad T_2 = \frac{81}{64}\, T_1$$

But $T_1 = 2.048$ kg-wt.

$$\therefore \quad T_2 = \frac{81}{64} \times 2.048 = 2.592 \text{ kg-wt.}$$

Dividing eq. (in) by (i), we have

$$\frac{9}{4} = \frac{4}{2}\sqrt{\left(\frac{T_3}{T_1}\right)}$$

or
$$\frac{81}{16} = \frac{16}{4}\left(\frac{T_3}{T_1}\right)$$

$$\therefore \quad T_3 = \frac{81 \times 4}{16 \times 16} T1$$

$$= \frac{81 \times 4 \times 2.048}{16 \times 16} = 2.592 \text{ kg-wt.}$$

Dividing eq. (iv) by (i), we have

$$\frac{27}{8} = 4\sqrt{\left(\frac{T_4}{T_1}\right)}$$

or
$$\frac{27 \times 27}{8 \times 8} = 16\frac{T_4}{T_1}.$$

$$\therefore \quad T_4 = \frac{27 \times 27}{8 \times 8 \times 16} T_1$$

$$= \frac{27 \times 27 \times 2.048}{8 \times 8 \times 16} = 1.458 \text{ kg-wt.}$$

12. A wire of length 60-0 cm and linear density 0.026 gm/cm is joined to another wire of length 86.6 cm and linear density 0.078 gm/cm. It is stretched between two supports fixed in a horizontal line under a tension of 10 kg. Find the lowest frequency of an external source which would set up standing waves in the compound wire with a node at the joint, and the total number of loops in which the wire would vibrate at this frequency.

Solution: The arrangement is shown in the diagram.

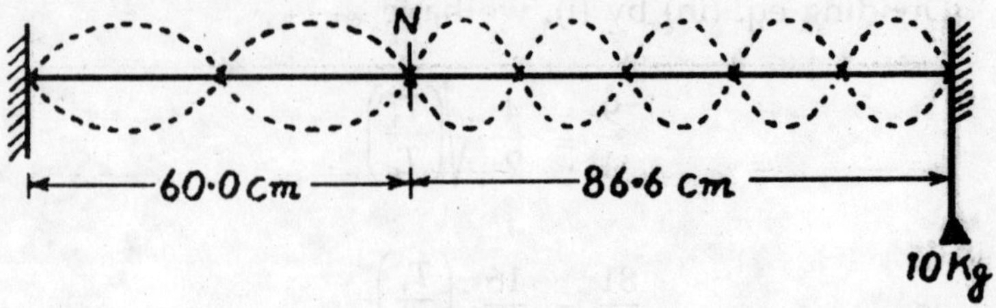

For the first wire; l=60.0 cm, m=0.026 gm/cm, and T= 10 × 1000 × 980 dynes. Therefore, its natural frequencies are

$$n = \frac{P}{2l}\sqrt{\left(\frac{T}{m}\right)}, \quad P = 1, 2, 3,....$$

$$= \frac{p}{2\times 60.0}\sqrt{\left(\frac{10\times 1000\times 980}{0.026}\right)}$$

$$= 161.7\ p = 161.7\times(1, 2, 3, 4,....)$$

$$= 161.7, 323.4, 485.1, 646.8,... \quad ...(i)$$

For the second wire; l = 86.6 cm, m = 0.078 gm/cm and T= 10 × 1000 × 980 dynes. Therefore, its natural frequencies are

$$n = \frac{P}{2l}\sqrt{\left(\frac{T}{m}\right)}, \quad P = 1, 2, 3,....$$

$$= \frac{p}{2\times 86.0}\sqrt{\left(\frac{10\times 1000\times 980}{0.078}\right)}$$

$$= 64.7\ p = 64.7\times(1, 2, 3, 4, 5, 6,...)$$

$$= 64.7, 129.4, 194.1, 256.8, 323.5, 388.2,... \quad ...(ii)$$

When the compound wire is excited, it would vibrate with a single frequency which must be one of the natural frequencies of both the individual wires if the standing waves are to be set up. We see from eq. (i) and (ii) that the natural frequency which is common to both wires is 323, which is the second harmonic of the first wire and fifth harmonic of the second wire. With this frequency the first wire would vibrate in 2 loops and the second one in 5 loops. Thus required frequency of the external source=323, and total number of loops in the compound wire=2+5=7.

13. *A fork and a string of length 100 cm give 4 beats per second. The string is made shorter, without any change of tension, until it is in unison with the fork. If its length is now 99 cm, what is the frequency of the fork?*

Solution: By the law of length of stretched strings, we have

$$n_1 l_1 = n_2 l_2.$$

On decreasing length, the frequency increases. Hence, if n be the frequency of the fork, then the initial frequency of the string is n–4 and the final frequency is n. Then, by the law of length, we have

$$(n-4) \times 100 = n \times 99$$

or $$100n - 400 = 99\, n$$

or $$n = 400.$$

14. *Two wires vibrate transversely in unison. If the tension in one is increased by 1 per cent, their simultaneous vibrations produce 3 beats every 2 seconds. Calculate their original frequency.*

Solution: Let T be the original tension in the wires. By the law of tension, the frequency of each is given by

$$n = k\sqrt{(T)}, \qquad \text{...(i)}$$

where k is a constant. When the tension of one is increased by 1 per cent, *i.e.* increased from T to $T + \dfrac{T}{100} = 1.01\ T$, the frequency increases from n to $n + \dfrac{3}{2}$ (since 3 beats are now heard in 2 seconds). Thus

$$n + \frac{3}{2} = k\sqrt{(1.01\,T)}$$

$$= 1.005\, k\sqrt{(T)}.$$

Substituting the value of $k\sqrt{(T)}$ from eq. (i), we get

$$n+\frac{3}{2} = 1.005\, n$$

or $$\frac{3}{2} = 1.005\, n - n = 0\text{-}005\, n.$$

$$\therefore \quad n = \frac{3}{2\times 0.005} = 300.$$

15. Two wires under the same tension vibrate with the same frequency 600. By what fraction the tension of one wire be increased to produce 6 beats per second when they vibrate together ?

Solution: Let T be the original tension in each wire. Since the frequency is proportional to the square-root of tension, we have

$$600 = k\sqrt{(T)}. \qquad \text{...(i)}$$

where k is a constant.

Let T' be the tension required in one of them so that its frequency becomes 606 to produce 6 beats with the other. Then

$$600 = k\sqrt{(T')}.$$

Squaring and dividing eq. (ii) by (i), we get

$$\frac{T'}{T} = \left(\frac{606}{600}\right)^2 = (1.01)^2 = 1.02.$$

16. The two parts of a sonometer wire divided by a movable knife-edge differ in length by 2 mm and produce 1 beat per second when sounded together. Find their frequencies, if the whole length of the wire is 1 metre.

Solution: Let l_1 and l_2 be the lengths of the two parts of the sonometer wire and n_1 and n_2 their frequencies. Then

$$l_1 + l_{2'} = 100 \text{ cm}$$

and $\quad l_1 - l_2 = 2.0$ mm=0.2 cm.

Solving these two equations, we get

$l_1 = 50.1$ cm $\quad$ and $\quad l_2 = 49.9$ cm.

According to the law of length of stretched strings, we have

$$n_1 l_1 = n_2 l_2.$$

But 1 beat is produced in 1 second, so that $n_1 - n_2 = 1$ or $n_2 = n_1 + 1$. Therefore

$$n_1 l_1 = (n_1 + 1) l_2.$$

sustituting the values of l_1 and l_2, we get

$$n_1 \times 50.1 = (n_1 + 1) \times 49.9$$

or $$50.1\, n_1 - 49.9\, n_1 = 49.9$$

or $$0.2\, n_1 = 49.9.$$

$\therefore$ $$n_1 = \frac{49.9}{0.2} = 249.5$$

and $$n_2 = n_1 + 1 = 249.5 + 1 = 250.5.$$

17. A tuning fork is found to give 4 beats per second when sounded in conjunction with a stretched string vibrating transversely under a tension of either 14.4 or 13.5 kilogram-weight. Calculate the frequency of the fork.

Solution: Let n be the frequency of the fork. Let n_1 and n_2 be the frequencies of the string when the tension is 14·4 and 13.5 kilogram-wt respectively. Since the frequency of a string is proportional to the square-root of the tension, we have

$$n_1 = k\sqrt{(14.4)}$$

$$n_2 = k\sqrt{(13.5)}$$

where k is a constant. In both cases the wire gives 4 beats per second with the fork of frequency n. Therefore, we have $n_1 - n = 4$ and $n_1 - n_2 = 4$, *i.e.*

$$k\sqrt{(14.4)} - n = 4, \qquad \text{...(i)}$$

and $$n - k\sqrt{(13.5)} = 4. \qquad \text{...(ii)}$$

Adding eq. (i) and (ii), we get

$$k\left[\left(\sqrt{(14.4)} - \sqrt{(13.5)}\right)\right] = 8$$

or $$k(3.795 - 3.675) = 8$$

$$\therefore \; k = \frac{8}{0.12}.$$

Substituting this value of k in eq. (i), we get

$$\frac{8}{0.12}\sqrt{(14.4)} - n = 4$$

or $$n = \frac{8 \times 3.795}{0.12} - 4 = 253 - 4 = 249.$$

18. In Melde's experiment the string vibrates in 5 loops with a load of 20 gnu How much load will be required to make the string vibrate in 2 loops ?

Solution: In Melde's experiment the string vibrates with the fixed frequency n of the fork. Hence if the length l is kept the same, the number of loops depends only upon the tension. If T_p be the tension required to produce p loops, we have

$$\text{n} = \frac{p}{2l}\sqrt{\left(\frac{T_p}{m}\right)}$$

or $$T_p \times p^2 = 4l^2mn^2 = \text{constant.}$$

Therefore, if T_5 and T_2 be the tensions to produce 5 and 2 loops respectively, we have

$$T_5 \times (5)^2 = T_2 \times (2)^2$$

or $$T_2 = \frac{T_2 \times (2)^2}{(2)^2} = \frac{25}{4}T_5.$$

Here, T_5=20 gm.

$$\therefore \quad T_2 = \frac{25}{4} \times 20 = 125 \text{ gm.}$$

19. The dimensions of the membrane of a rectangular drum are 10 cm x 20 cm. It is stretched to a tension of 5 kg/cm and its mass is 20 gm. What is the fundamental frequency?

Solution: The fundamental frequency of a rectangular membrane of sides a and b is given by

$$n_{1,1} = \frac{1}{2}\sqrt{\left[\frac{T}{m}\left(\frac{1}{a^2}+\frac{1}{b^2}\right)\right]},$$

where T is the tension per unit length of the boundary of the membrane and m is the mass per unit area of the membrane.

Here, T=5 kg/cm=5×1000×980 dyne/cm m=20/(10×20) =0.1 gm/cm², a=10 cm and b=20 cm.

$$\therefore \quad n_{1,1} = \frac{1}{2}\sqrt{\left[\frac{5\times 1000\times 980}{0.1}\left(\frac{1}{10^2}+\frac{1}{20^2}\right)\right]}$$

$$= \frac{1}{2}\sqrt{\left[49\times 10^6\left(\frac{4+1}{400}\right)\right]}$$

$$= 391 \text{ per sec.}$$

20. A square membrane is made of material of density m kg/m³ and is under a tension of T N/m. What should be the size of each side of the membrane in order that the fundamental frequency be n?

Solution: The fundamental frequency of a square membrane of side a is given by

$$n = \frac{1}{2}\sqrt{\left[\frac{T}{m}\left(\frac{1}{a^2}+\frac{1}{b^2}\right)\right]} = \frac{1}{2}a\sqrt{\left(\frac{2T}{m}\right)} = \frac{1}{a}\sqrt{\left(\frac{2T}{m}\right)}$$

or $$a = \frac{1}{n}\sqrt{\left(\frac{T}{m}\right)} \text{ metre.}$$

21. Compute the lowest degenerate modes of vibration of a rectangular membrane of sides a and 2a.

Solution: The frequency of a mode (h, k) is given by

$$n_{h,k} = \frac{1}{2}\sqrt{\left[\frac{T}{m}\left(\frac{h^2}{a^2}+\frac{k^2}{b^2}\right)\right]},$$

where a and b are the sides of the membrane. Now, two modes (h_1, k_1) and (h_2, k_2) are degenerate when

$$n_{h1, k1} = n_{h2, k2} \quad \text{(where } h_1 \neq h_2,\ k_1 \neq k_2\text{)}$$

or $$\frac{h_1^{\ 2}}{a^2} + \frac{k_1^{\ 2}}{b^2} = \frac{k_2^{\ 2}}{b^2} + \frac{h_2^{\ 2}}{a^2}.$$

This is true only when

$$\frac{h_1^{\ 2}}{a^2} = \frac{k_2^{\ 2}}{b^2} \text{ and } \frac{k_1^{\ 2}}{b^2} = \frac{h_2^{\ 2}}{a^2}$$

or $$\frac{h_1}{a} = \frac{k_2}{b} \text{ and } \frac{k_1}{b} = \frac{h_2}{a}$$

or $$h_1 = \left(\frac{a}{b}\right) k_2 \text{ and } k_1 = \left(\frac{b}{a}\right) h_2.$$

In the given problem, $b=2a$. Hence the above conditions become

$$h_1 = \frac{1}{2} k_2 \text{ and } k_1 = 2\, h_2.$$

If $h_1=\alpha$ then $k_2=2\alpha$; and if $h_2=\beta$, $k_1=2\beta$ (where α and β are integers). Thus the two normal degenerate modes are

$$(\alpha, 2\beta) \text{ and } (\beta, 2\alpha).$$

The lowest degenerate modes are obtained by putting $\alpha=1$ and $\beta=2$. Thus they are (1, 4) and (2, 2).

22. *Find oat the lowest degenerate modes of vibration of a rectangular membrane of sides 6 and 3 units.*

[Ans. (2, 2) and (4, 1)]

7

Waves in Dispersive and Absorptive Media

Wave Velocity

When a monochromatic wave (wave of a single frequency or wavelength) travels through a medium, its velocity of advancement in the medium is called the 'wave velocity'. For example, a plane harmonic wave travelling along the $+x$ direction is given by

$$y = a \sin. (\omega t - kx),$$

where a is the amplitude, ω $(=2\pi n)$ is the angular frequency and k $(=2\pi/\lambda)$ is the propagation constant of the wave. By definition, the ratio of angular frequency ω to the propagation constant k is the wave velocity v. That is,

$$\boxed{v = \frac{\omega}{k}}$$

$(\omega t - kx)$ is the phase of the wave-motion. Therefore, the planes of constant phase (wave fronts) are defined by

$$\omega t - kx = \text{constant}.$$

Differentiating with respect to time, we get

$$\omega - k\frac{dx}{dt} = 0.$$

$$\frac{dx}{dt} = \frac{\omega}{k},$$

which is the wave velocity v. Thus the wave velocity is the velocity with which planes of constant phase advance through the medium. Hence the wave velocity is also called as 'phase velocity'.

Group Velocity: In practice, we come across pulses rather than monochromatic waves. A pulse consists of a number of waves differing slightly from one another in frequency. A superposition of these waves is called a 'wave packet' or a 'wave group'.

When such a group travels in a dispersive medium, the phase velocities of its different components are different. The observed velocity is, however, the velocity with which the maximum amplitude of the group advances in the medium. This is called the 'group velocity'. Thus the group velocity is the velocity with which the energy in the group is transmitted. The individual waves travel "inside" the group with their phase velocities.

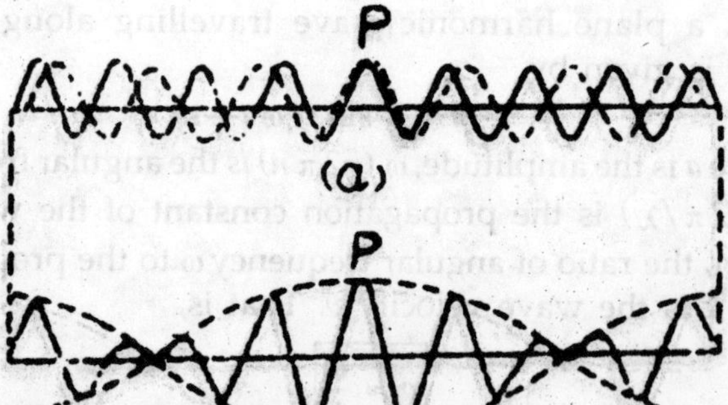

In the figure above are shown two plane harmonic waves of equal amplitudes but slightly different frequencies travelling from left to right. The dotted curve represents the wave of lower frequency and is travelling faster.

At a certain instant the two waves are in phase at the point *P*. Therefore, at this instant, the maximum amplitude of the group formed by them also lies at *P*. At a later instant the maximum will be build up a little to the left of *P*. That is, the maximum will move to the left with time relative to the waves. As a result, the group velocity will be lower than the wave velocities.

Expression for Group Velocity: Let us consider a wave group consisting of two components of equal amplitudes *a* but slightly different angular frequencies ω_1 and ω_2, and propagation constants k_1 and k_2. Their separate displacements are given by

$$y_1 = a \sin(\omega_1 t - k_1 x), \qquad \text{...(i)}$$

and $$y_2 = a \sin(\omega_2 t - k_2 x). \qquad \text{...(ii)}$$

Their superposition gives

$$y = y_1 + y_2$$

$$= a\,[\sin(\omega_1 t - k_1 x) + \sin(\omega_2 t - k_2 x)].$$

Using the trigonometric relation, $\sin A + \sin B = 2 \sin\dfrac{A+B}{2} \cos\dfrac{A-B}{2}$, We get

$$y = 2\,a \sin\left[\frac{(\omega_1+\omega_2)t}{2} - \frac{(k_1+k_2)x}{2}\right] \cos\left[\frac{(\omega_1-\omega_2)t}{2} - \frac{(k_1-k_2)x}{2}\right]$$

or $$y = 2\,a \cos\left[\frac{(\omega_1-\omega_2)t}{2} - \frac{(k_1-k_2)x}{2}\right] \sin\left[\frac{(\omega_1+\omega_2)t}{2} - \frac{(k_1+k_2)x}{2}\right]$$

This represents a wave-system with a frequency$(\omega_1 t + \omega_2)/2$ which is very close to the frequency of either component, but with an amplitude given by

$$A = 2\,a \cos\left[\frac{(\omega_1-\omega_2)t}{2} - \frac{(k_1-k_2)x}{2}\right].$$

Thus the amplitude of the wave group is modulated both in space and time by a very-slowly varying envelope of frequency $(\omega_1 - \omega_2)/2$ and propagation constant $(\kappa_1 - k_2)/2$, and has a maximum value of *2a*. This envelope is represented by the dotted curve in figure. The velocity with which this envelope moves, that is, the velocity of the maximum amplitude of the group is given by

$$v_g = \frac{\omega_1 - \omega_2}{k_1 - k_2} = \frac{\Delta\omega}{\Delta k}$$

If a group contains a number of frequency-components in an infinitely small frequency-interval, then the expression for the group velocity may be written as

$$\boxed{v_g = \frac{d\omega}{dk}}$$

Relation between Group Velocity and Wave Velocity: Since $\omega = kv$, where v is wave (phase) velocity, the group velocity is given by

$$v_g = \frac{d\omega}{dk} = \frac{d}{dk}(kv)$$

$$= v + k\frac{dv}{dk}$$

Now, $k = \frac{2\pi}{\lambda}$, where λ is wavelength. Therefore

$$v_g = v + \frac{2\pi}{\lambda}\frac{dv}{d\left(\frac{2\pi}{\lambda}\right)}$$

$$= v + \frac{1}{\lambda}\frac{dv}{d\left(\frac{1}{\lambda}\right)}$$

But $d\left(\frac{1}{\lambda}\right) = -\frac{1}{\lambda^2}d\lambda$. Therefore

$$\boxed{v_g = v - \lambda\frac{dv}{d\lambda}}$$

This is the relation between group velocity v_g and wave velocity v in a dispersive medium.

Normal and Anomalous Dispersion: Usually dv/dX is positive, that the group velocity v_g is smaller than the wave velocity v. This is called 'normal dispersion'. But 'anomalous dispersion' arise when $dv/d\lambda$ is negative, so that v_g is greater than v. An electrical conductor is anomalously dispersive to electromagnetic waves.

Non-dispersive Medium: In a non-dispersive medium ($v = \omega / k$ = constant), we have $\frac{dv}{d\lambda} = 0$, so that

$$v_g = v_0$$

that is, the two velocities are identical. For light waves in free space the group velocity is same as the wave velocity,

Sound waves in gases are also non-dispersive. This is a fortunate circumstance. If sounds of different frequencies travelled at different speeds through the air it would result in chaos and aural anguish. We could never listen to an orchestra.

Group Velocity of Gravity Waves

The phase velocity of waves in deep water (gravity waves) is given by

$$v = \sqrt{\left(\frac{g\lambda}{2\pi}\right)}$$

Let us write $\frac{2\pi}{\lambda} = k$, where k is propagation constant. Then the phase or wave velocity, defined by ω / k, is given by

$$v = \frac{\omega}{k} = \sqrt{\left(\frac{g}{k}\right)} \qquad \text{.....(i)}$$

or

$$\omega = \sqrt{(gk)}.$$

Now, the group velocity, defined by $d\omega/dk$, is given by

$$v_g = \frac{d\omega}{dk}$$

$$= \frac{d}{dk}(gk)^{1/2}$$

$$= (g^{1/2})\ \tfrac{1}{2}k^{-1/2}$$

$$= \tfrac{1}{2}\sqrt{\left(\frac{g}{k}\right)}$$

But $\sqrt{\left(\frac{g}{k}\right)} = v$ by eq.(i) . Hence

$$v_g = \tfrac{1}{2}v,$$

This is what we had to prove.

The wave velocity for $\lambda = 1000$ metre is,

$$v = \sqrt{\left(\frac{g\lambda}{2\pi}\right)}$$

$$= \sqrt{\left(\frac{9.8\times1000}{2\times3.14}\right)} = 39.5 \text{ metre/sec.}$$

The group velocity is

$$v_g = \frac{v}{2} = 19.75 \text{ metre/sec.}$$

Thus, the gravity waves are highly dispersive. The individual waves travel much more rapidly within the group.

Group Velocity of Surface Waves

The wave (phase) velocity on water surface is controlled mainly by surface tension and is given by

$$v = \sqrt{\left(\frac{2\pi T}{\lambda\rho} + \frac{g\lambda}{2\pi}\right)}$$

Let us write $\frac{2\pi}{\lambda} = k$, where k is propagation constant. Then the wave velocity, defined by ω/k, is given by

$$v = \frac{\omega}{k} = \sqrt{\left(\frac{kT}{\rho} + \frac{g}{k}\right)} \qquad \text{...(i)}$$

or $$\omega = \sqrt{\left(\frac{k^3T}{\rho} + gk\right)}$$

Now, the group velocity, defined by $d\omega/dk$, is given by

$$v_g = \frac{d\omega}{dk}$$

$$= \frac{d\omega}{dk}\left(\frac{k^3T}{\rho} + kg\right)^{1/2}$$

$$= \tfrac{1}{2}\left(\frac{k^3T}{\rho} + kg\right)^{-1/2}\left(\frac{3k^2T}{\rho} + g\right)$$

$$= \tfrac{1}{2}k^{-1}\left(\frac{k^2T}{\rho} + \frac{g}{k}\right)^{-1/2}\left(\frac{3kT}{\rho} + \frac{g}{k}\right)$$

$$= \tfrac{1}{2}\left(\frac{kT}{\rho} + \frac{g}{k}\right)^{-1/2}\left(\frac{3kT}{\rho} + \frac{g}{k}\right)$$

$$= \tfrac{1}{2}v^{-1}\left(\frac{3kT}{\rho} + \frac{g}{k}\right) \qquad \text{[by eq. (i)]}$$

$$= \tfrac{2}{3}\left(\frac{kT}{\rho}\right)\frac{1}{v} + \frac{1}{2}\left(\frac{g}{k}\right)\frac{1}{v}.$$

This is the expression for the group velocity v_g in terms of the wave velocity v.

We can see that if we neglect the effect of gravity (g=0), then $v_g = \tfrac{3}{2}v$. On the other hand if we neglect the effect of surface tension $(T=0)$, then $v_g = \tfrac{1}{2}v$

Group Velocity of Matter Waves

The angular frequency is

$$\omega = 2\pi n.$$

Here the wave frequency is $n = (1/h)\tfrac{1}{2}mv^2$.

$$\therefore \qquad \omega = \frac{2\pi m v^2}{2h}.$$

Differentiating with respect to v, we get

$$\frac{d\omega}{dv} = \frac{2\pi m v}{h} \qquad ...(i)$$

Again, the propagation constant is

$$k = \frac{2\pi}{\lambda}$$

For matter wave, $\lambda = h/mv$.

$$\therefore \qquad k = \frac{2\pi}{\lambda}$$

$$\text{or} \qquad \frac{dk}{dv} = \frac{2\pi m}{h} \qquad ...(ii)$$

Dividing eq. (i) and (ii), we have

$$\frac{d\omega}{dk} = v.$$

By definition, $\frac{d\omega}{dk}$ is the group velocity v_g.

$$\therefore \qquad v_g = v.$$

Product of Group Velocity and Phase Velocity is a Constant

The given relation is

$$\omega^2 = c^2k^2 + \omega_0^2$$

Differentiating ω with respect to k, we get

$$2 = \frac{d\omega}{dk} c^2 (2k)$$

$$\text{or} \qquad \frac{d\omega}{dk} = \frac{c^2 k}{\omega}$$

By definition, $\frac{d\omega}{dk}$ is the group velocity v_g.

$$\therefore \qquad v_g = \frac{c^2 k}{\omega} = \frac{c^2}{\omega / k}$$

Again, by definition, ω / k is the phase velocity v.

$$\therefore \qquad v_g = \frac{c^2}{v}$$

or $$v_g \times v = c^2 \text{ (constant)}$$

Superposition of Many Harmonic Waves in a Narrow Frequency-range $\omega\ \Delta$

The Band-width Theorem: The wave group having only two wave components may be extended to a group of many wave components differing slightly from one another in frequency.

Let us consider a group (or pulse) of n frequency components, each of amplitude a, lying within a narrow frequency range $\Delta\ \omega$. Let $d\omega$ be the constant frequency difference between successive components. The constant phase difference between successive components arising from the frequency difference is $(d\omega)t$.

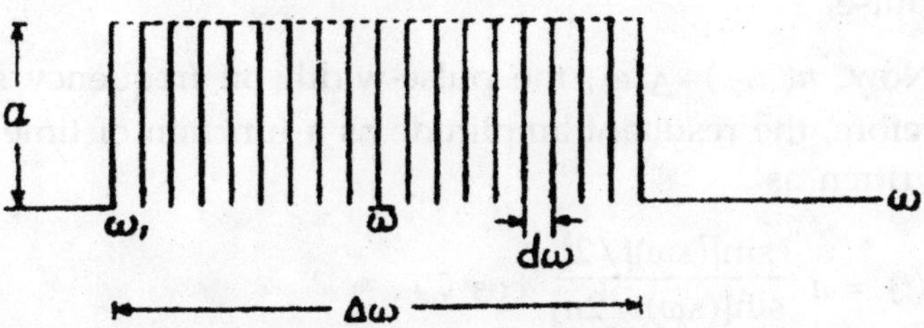

The amplitude which results from the superposition of the frequency components is given by

$$R = a \cos \omega_1 t + a \cos(\omega_1 + d\omega)\, t + a \cos(\omega_1 + 2\, d\omega)\, t + \ldots + a \cos\{\omega_1 + (n-1)\, d\omega\}\, t.$$

Let us express this series as a geometric progression in complex exponential notation :

$$R = ae^{j\omega_1 t} + ae^{j(\omega_1 + d\omega)t} + ae^{j(\omega_1 + 2d\omega)t} + \ldots + ae^{j\{\omega_1 + (n-1)(d\omega)t}$$

$$= ae^{j\omega_1 t}[1+e^{j(d\omega)t}+e^{j2(d\omega)t}+...+e^{j(n-1)(d\omega)t}]$$

$$= ae^{j\omega_1 t}\frac{I-e^{jn(d\omega)t}}{I-e^{j(d\omega)t}}$$

$$= ae^{j\omega_1 t}\frac{e^{jn(d\omega)^{t/2}}(e^{-jn(d\omega)t/2}-e^{jn(d\omega)t/2})}{e^{j(d\omega)^{t/2}}(e^{-j(d\omega)^{t/2}}-e^{j(d\omega)^{t/2}})}$$

$$= ae^{j\omega_1 t}e^{j(n-1)(d\omega)^{t/2}}\frac{\sin[n(d\omega)t/2]}{\sin[(d\omega)t/2]}$$

Recovering the original cos term from the complex exponential notation, we get the resultant amplitude given by

$$R = a\cos\left[\{\omega_1+\tfrac{1}{2}(n-1)\ (d\omega)\}t\right]\frac{\sin[n(d\omega)t/2]}{\sin[(d\omega)t/2]}$$

$$= a\ \frac{\sin[n(d\omega)t/2]}{\sin[(d\omega)t/2]}\ \cos\ \overline{\omega}t$$

where $\overline{\omega}t\ \{=\omega_1+\frac{1}{2}(n-1)\ (d\omega)\}$ is the average frequency in the pulse.

Now, $n(d\omega)=\Delta\ \omega$, the pulse-width on frequency scale. Therefore, the resultant amplitude as a function of time may be written as

$$R(t)\ = a\ \frac{\sin[(\Delta\omega)t/2]}{\sin[(\Delta\omega)t/2n]}\ \cos\ \overline{\omega}t\cdot$$

When n is large, we may take $\sin\frac{(\Delta\omega)t}{2n}=\frac{(\Delta\omega)t}{2n}$, so that

$$R(t)\ = na\ \frac{\sin[(\Delta\omega)t/2]}{(\Delta\omega)t/2n}\ \cos\ \overline{\omega}t\cdot$$

Let us put $na=A$ and $(\Delta\ \omega)t/2$—α, which is half the phase difference between the first and the last component at time t. Then, we have

$$R(t) \doteq A \frac{\sin \alpha}{\alpha} \cos \bar{\omega} t \cdot$$

This expression gives the time behaviour of the pulse.

At $t=0$, $\alpha=0$, so that $\frac{\sin \alpha}{\alpha} \rightarrow$ I, and $\cos \bar{\omega} t = 1$: so we have $R(t) = A = na$.

That is, at $t=0$, all the components superpose with zero phase difference to give the maximum amplitude $R(t)=na$. After a time-interval ΔT when

$$\alpha = \frac{\Delta\omega \Delta T}{2} = \pi$$

$$\frac{\sin \alpha}{\alpha} = 0, \text{ and so}$$

$$R(t) = 0.$$

That is, at $t=\Delta T$, the phases between the frequency components are such that the resulting amplitude $R(t)$ is zero.

The time-interval ΔT which is a measure of the width of the pulse on time scale is therefore given by

$$\frac{\Delta\omega \Delta T}{2} = \pi$$

so that
$$\Delta\omega = \frac{2\pi}{\Delta T} = 2\pi\Delta\nu.$$

This is the required expression for the band-width.

Attenuation of Waves in Absorptive Medium

When waves propagate through a medium, the particles comprising the medium are set into oscillation. The relative motion between the various particles is opposed by damping forces, such as the viscous forces.

Therefore, the propagating wave suffers energy loss (attenuation) against the damping forces. If this loss is not too serious, the propagation of the wave can be described by combining wave equation and diffusion equation.

The wave equation is

$$\frac{\delta^2 y}{\delta x^2} = \frac{1}{v^2}\frac{\delta^2 y}{\delta t^2}$$

and the diffusion equation is

$$\frac{\partial^2 y}{\partial x^2} = \frac{1}{d}\frac{\delta y}{\delta t}.$$

The constant d, called the diffusivity, takes the value η / ρ (coeff. of viscosity/density) is a viscous medium.

The wave equation in a diffusing medium is therefore

$$\frac{\partial^2 y}{\partial x^2} = \frac{1}{v^2}\frac{\partial^2 y}{\partial t^2} + \frac{1}{d}\frac{\partial y}{\partial t} \quad \text{...(i)}$$

Let the solution to this equation be

$$y = y_m e^{j(\omega t - \gamma x)} \quad \text{...(ii)}$$

where y_m is the maximum amplitude and $j = \sqrt{(-1)}$. Differentiating eq. (ii) partially with respect to x and t, we have

$$\frac{\partial y}{\partial x} = y_m e^{j(\omega t - \gamma x)}(-j\gamma),$$

$$\frac{\partial^2 y}{\partial x^2} = y_m e^{j(\omega t - \gamma x)}(-j\gamma)^2 = -\gamma^2 y,$$

$$\frac{\partial y}{\partial t} = y_m e^{j(\omega t - \gamma x)}(j\omega) = j\omega y$$

and $$\frac{\partial^2 y}{\partial t^2} = y_m e^{j(\omega t - \gamma x)}(j^2\omega^2) = -\omega^2 y.$$

Substituting the values of $\frac{\partial^2 y}{\partial x^2}, \frac{\partial^2 y}{\partial t^2}$ and $\frac{\partial y}{\partial t}$ in eq. (i), we get

$$-\gamma^2 y = \frac{1}{v^2}(-\omega^2 y) + \frac{1}{d}(j\omega y)$$

or $$\gamma^2 = \frac{\omega^2}{v^2} - j\frac{\omega}{d}.$$

Thus γ has a complex value. We know that the velocity is defined by

$$v = \frac{\omega}{k},$$

where k is propagation constant. Substituting this value of v in eq. (iii), we get

$$\gamma^2 = k^2 - j\frac{\omega}{d} \qquad \text{...(iii)}$$

Let us put $\gamma = k - j\alpha$. Then we obtain

$$\gamma^2 = k^2 - \alpha^2 - 2jk\alpha$$

$$\simeq k^2 - 2jk\alpha \quad (\text{if } \alpha << k) \qquad \text{...(iv)}$$

Comparing the imaginary parts of eq. (iii) and (iv), we get

$$\frac{\omega}{d} = 2\,k\alpha$$

or $$\alpha = \frac{\omega}{2kd} = \frac{v}{2d}$$

The solution for y, eq. (ii), then becomes

$$y = y_m\, e^{j(\omega t - \gamma x)}$$

$$= y_m\, e^{j[\omega t - (k - j\alpha)x]}$$

$$= y_m\, e^{j[(\omega t - kx) + j\alpha x]}$$

$$= y_m e^{-\alpha x} e^{j}(\omega t - kx)$$

This shows that the wave is a sine or a cosine oscillation of amplitude $y_m e^{-\alpha x}$ which decays exponentially with time.

The physical significance of the condition $\alpha << \kappa = 2\pi/\lambda$ is that many wavelengths λ are contained in the distance $1/\alpha$ before the amplitude decays to y_m/e at $x = 1/\alpha$.

Since the amplitude decays with distance as $e^{-\alpha x}$, the energy in the wave being proportional to the square of the amplitude decays as $e^{-2\alpha x}$ or $e^{-\beta x}$, where $\beta = 2\alpha$.

Problems

1. The refractive indices of a medium for two spectral lines of wavelengths 4000Å and 5000Å are 1.540 and 1.530 respectively. Calculate the value of the group velocity. Velocity of light in vacuum is 3×10^{10} cm/sec.

Solution: The group velocity v_g is defined by

$$v_g = \frac{\Delta\omega}{\Delta k} = \frac{\omega_2 - \omega_1}{k_2 - k_1} \qquad \text{...(i)}$$

where ω is angular frequency and k is propagation constant.

Now $\omega = 2\pi n$,

where n is frequency, which is given by

$$n = \frac{c}{\lambda_{vac}}$$

where λ_{vac} is the wavelength of light in vacuum. If λ be the wavelength in a medium of refractive index μ then

$$\lambda = \lambda_{vac}/\mu$$

and so
$$n = \frac{c}{\mu\lambda}$$

Thus
$$\omega = 2\pi\frac{c}{\mu\lambda}$$

Also,
$$k = \frac{2\pi}{\lambda}$$

From eq. (i), we have

$$v_g = \frac{2\pi c\left(\dfrac{1}{\mu_2\lambda_2} - \dfrac{1}{\mu_1\lambda_1}\right)}{2\pi\left(\dfrac{1}{\lambda_2} - \dfrac{1}{\lambda_1}\right)}$$

$$= \frac{c(\mu_1\lambda_1 - \mu_2\lambda_2)}{\mu_1\mu_2(\lambda_1 - \lambda_2)}$$

Given: $c=3\times10^{10}$ cm/sec, μ_1=1.540, μ_2=1.530, λ_1= 4000Å, λ_2= 5000Å,

$$\therefore v_g = \frac{3\times10^{10}(1.540\times4000 - 1.530\times5000)}{1.540\times1.530(4000-5000)} = 1.897\times1010\text{cm/sec.}$$

2. *A wave packet of mean wavelength 3.6×10^{-5} cm is travelling with phase velocity 1.8×10^{10} cm/sec in a dispersive medium of carbon-disulphide. From the dispersion of the medium, $\frac{dv}{d\lambda}$ is computed to be 3.8×10^{18} cm per sec per cm. Calculate the group velocity in carbon-disulphide.*

Solution: The group velocity v_g is related to the phase velocity by

$$v_g = v - \lambda\frac{dv}{d\lambda}$$

Substituting the given values, we get

$$\begin{aligned} v_g &= (1.8\times10^{10}\text{ cm/sec})\ (3.6\times10^{-5}\text{cm})(3.8\times10^{13}/\text{sec}) \\ &= 1.8\times10^{10}\text{ cm/sec}-13.68\times10^{8}\text{ cm/sec} \\ &= 1.8\times10^{10}\text{ cm/sec}-0.1368\times10^{10}\text{cm/sec} \\ &= 1.66\times10^{10}\text{ cm/sec.} \end{aligned}$$

3.*(i) Calculate the wave number for gamma rays of wavelength 1Å.*

(ii) The angular frequency is five times of the propagation constant. Calculate phase and group velocities.

Solution: (i) The wave number v (say) is defined by

$$v = \frac{1}{\lambda}$$

For the given gamma rays, λ =1Å = 10^{-8} cm.

$$\therefore \qquad v = \frac{1}{\lambda} = \frac{1}{10^{-8}\, cm} = 10^{8} \text{ per cm.}$$

(ii) The angular frequency ω is 5 times the propagation constant, that is,

$$\frac{\omega}{k} = 5$$

$\frac{\omega}{k}$ is the phase velocity v. Thus

$$v = 5 \text{ cm/sec,}$$

if k is in per cm.

Since, here ω/k is constant, that is,

$$v = \text{constant}$$

$$\therefore \qquad \frac{dv}{d\lambda} = 0.$$

Now, the group velocity is given by

$$v_g = v - \lambda \frac{dv}{d\lambda} = 0.$$

because $\frac{dv}{d\lambda} = 0$. Thus

$$v_g = 5 \text{ cm, sec.}$$

4. *The intensity of a plane wave is reduced to half after travelling one metre in an absorbing medium. What will be its after travelling three metres? What will happen if the is non-absorbing?*

Solution: Let I_0 be the initial intensity of the wave, and I the intensity remaining after the wave travels a distance x in an absorbing medium. Then, we have

where β is a constant.

$$I = I_0 e^{-\beta x},$$

When x=1 metre, $I = \frac{1}{2} I_0$. Thus

$$\tfrac{1}{2} = e^{-\beta}$$

or $$\beta = \log_e 2 = 0.693.$$

Let I be the intensity for $x = 3$ metres. Thus

$$I = I_0 e^{-8\beta}$$

But $$3\beta = 3 \times 0.693 = 2.08.$$

$\therefore$ $$I = I_0 e^{-2.08} = \frac{I_0}{e^{2.08}}$$

We can see that $e^{2.08} = 8$.

$\therefore$ $$I = \frac{1}{8} I_0.$$

The given wave is a "plane wave." Hence while travelling in a 'non-absorbing' medium, the intensity will remain undiminished.

5. *The intensity I_0 of a plane wave is reduced by 30 per cent after passing through 1 metre in an absorbing medium. What distance should it travel so that the intensity becomes approximately $I_0/2$?*

Solution: For a plane wave, the variation of intensity with distance in an absorbing medium is given by an exponential law of the form

$$I = I_0 e^{-\beta x}.$$

where I_0 is the initial intensity, I is the intensity after travelling a distance x and β is a constant.

When $x = 1$ metre, $I = 0.7\ I_0$. Thus

$$0.70\ I_0 = I_0 e^{-\beta}$$

or $$\frac{1}{0.70} = e^{\beta}$$

or $$\beta = \log_e (10/7) = \log_e (1.428)$$

For $I = \tfrac{1}{2} I_0$, we have

or $$\tfrac{1}{2} I_0 = I_0 e^{-\beta x}$$

or $$2 = e^{-\beta x}$$

$$\beta x = \log_e 2$$

or $$x = \frac{\log_e 2}{\beta} = \frac{\log_e 2}{\log_e 1.428}$$

$$= \frac{\log_{10} 2}{\log_{10} 1.428} = \frac{0.3010}{0.1549} \simeq 2 \text{ metre.}$$

8

Velocity of Acoustic Waves

Velocity of Longitudinal Waves

When a metallic rod is excited longitudinally by stroking it lengthwise with a resined cloth, its cross-sections oscillate to and fro in the direction of the length of the rod and longitudinal waves are set up in it. The velocity of these waves depends upon the density and elastic modulus of the material of the rod.

Let us consider a solid rod whose length is taken along the Af-axis. Let there be two plane sections P, Q of this rod at distances x and $x+\delta x$ respectively from the origin.

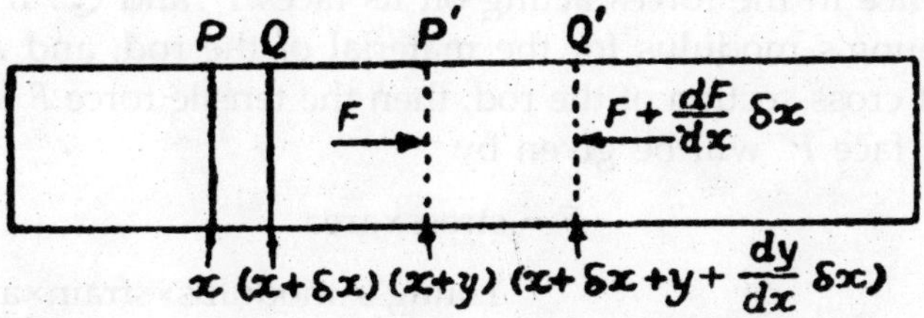

Suppose at some particular instant, when the longitudinal wave is passing, the displaced positions of the original planes

P and Q are P' and Q' respectively. Let the displacement of the plane P be y.

Since the rate of change of displacement with distance is dy/dx, the displacement of the plene Q will be $\left(y+\frac{dt}{dx}\delta x\right)$. Thus the new positions of the planes, P' and Q', are $x+y$ and $\left(x+\delta x+y+\frac{dy}{dx}\delta x\right)$ respectively.

Now, the initial length of the slice between the planes P and Q

$$= (x+\delta x) - x = \delta x,$$

and the final length of the slice between the planes P' and Q'

$$=\left(x+\delta x+y+\frac{dy}{dx}\delta x\right) - (x+y) = \left(\delta x+\frac{dy}{dx}\delta x\right).$$

$\therefore$ increase in length of the slice

$$=\left(\delta x+\frac{dy}{dx}\delta x\right)-\delta x=\frac{dy}{dx}\delta x.$$

$$\therefore \text{ longitudinal strain} = \frac{\text{increase in length}}{\text{initial length}}$$

$$\frac{\frac{dy}{dx}\delta x}{\delta x} = \frac{dy}{dx}.$$

As the slice is under longitudinal strain, there must be a difference in the forces acting on its faces P' and Q'. If Y be the Young's modulus for the material of the rod, and A the area of cross-section of the rod, then the tensile force F acting on the face P' will be given by

$$F = \text{stress} \times \text{area}$$

$$= \text{Young's modulus} \times \text{strain} \times \text{area}$$

$$= \text{y} \times \frac{dy}{dx} \times A. \qquad \text{.....(i)}$$

The force acting on the face Q' of the slice will be in the opposite sense and equal to $F+\frac{dF}{dx}\delta x$.

$\therefore$ resultant force acting on the slice

$$= F+\frac{dF}{dx}\delta x - F = \frac{dF}{dx}\delta x$$

$$= \frac{d}{dx}\left(y\frac{dy}{dx}A\right)\delta x \qquad \text{[from eq.(i)]}$$

$$= y\,A\left(\frac{d^2y}{dx^2}\right)\delta x.$$

If ρ be the density of the material of the rod, the mass of the slice between the planes will be $\rho A\ \delta x$. Then, since force=mass $\times$ acceleration, the resultant force acting on the slice must be $\rho A\,\delta x\left(\frac{d^2y}{dx^2}\right)$, where $\frac{d^2y}{dt^2}$ is the instantaneous acceleration. Hence, the equation of motion is

$$YA\left(\frac{d^2y}{dx^2}\right)\delta x = \rho A\,\delta x\left(\frac{d^2y}{dx^2}\right)$$

$$\frac{d^2y}{dt^2} = \frac{Y}{\rho}\frac{d^2y}{dx^2}$$

Comparing it with the differential equation of wave motion

$$\frac{d^2y}{dt^2} = v^2\frac{d^2y}{dx^2}, \text{ we get}$$

$$v^2 = \frac{Y}{\rho}$$

$$v = \sqrt{\frac{Y}{\rho}}$$

This is the expression for the velocity of longitudinal waves in a rod. The velocity is proportional to the square-root of the

Young's modulus and inversely proportional to the square-root of the density of the solid material.

If the solid medium is extended, the velocity of the longitudinal waves will depend on the bulk (volume) and shape elasticity also.

Frequency of a Rod Vibrating Longitudinally: Let a rod of length l, clamped at two points distant l4 from ends, be excited longitudinally.

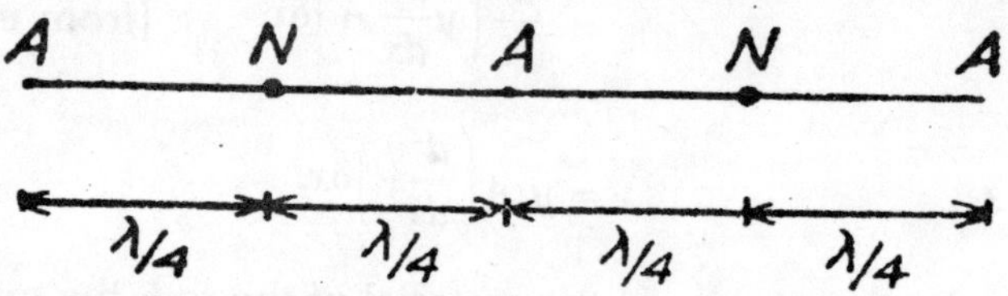

These two points will be nodes. The free ends and the middle point will be antinodes. Since the distance between an antinode and the nearest node is $\lambda/4$, the length of the rod is

$$l = \frac{\lambda}{4} + \frac{\lambda}{4} + \frac{\lambda}{4} + \frac{\lambda}{4} = \lambda .$$

If n be the frequency of vibration

$$v = n\lambda = nl$$

$$\therefore \qquad n = \frac{v}{l}$$

But

$$v = \sqrt{\frac{Y}{\rho}}$$

$$\therefore \qquad n = \frac{1}{l}\sqrt{\left(\frac{Y}{\rho}\right)}$$

Velocity Sound

Only longitudinal waves can travel in gases. Whenever such waves (say sound waves) travel in a gas, the particles of the gas oscillate in the direction of wave propagation. The phase of oscillation varies from particle to particle. The distances between the particles so change that at any instant the particles are alternately crowded together and spread out. Therefore, the pressure varies from particle to particle in the gas.

Let us consider a tube of air of unit cross-section, with its length along the axis of x. Let A and B be two planes at right angles to the axis of the tube at distances x and $x+\delta x$ respectively from the origin.

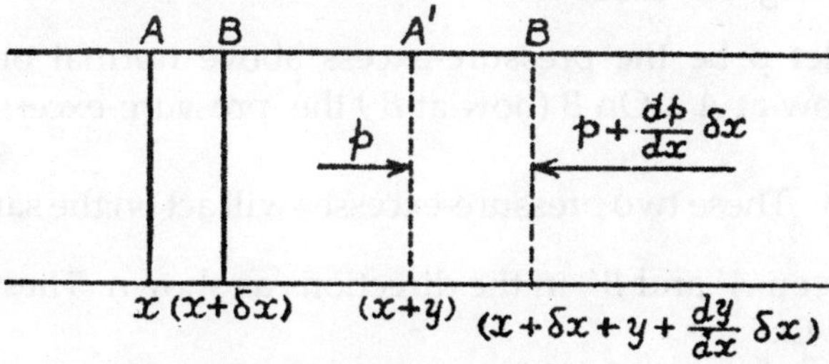

Let a plane sound wave pass along the axis of the tube. Suppose, at some particular instant when the wave is passing, the displaced positions of the planes are A' and B' respectively. Let the displacement of the plane A be y. Since the rate of change of displacement with distance is dy/dx, the displacement of the plane B will be $\left(y+\frac{dt}{dx}\delta x\right)$. Thus the new positions of the planes, A' and B', are $(x+y)$ and $\left(x+\delta x+y+\frac{dy}{dx}\delta x\right)$ respectively.

The area of cross-section of the tube considered is unity. Therefore the initial volume of air between the planes (A and B)

$$= (x+\delta x) - x = \delta x$$

The final volume of air between the planes (A' and B')

$$\left(x+\delta x+y+\frac{dy}{dx}\delta x\right) - (x+y) = \left(\delta x+\frac{dt}{dx}\delta x\right). \therefore$$

increase in volume

$$= \left(\delta x+\frac{dt}{dx}\delta x\right) - -\delta x = \frac{dy}{dx}\delta x$$

$$\therefore \quad \text{volume strain} = \frac{\text{increase in volume}}{\text{initial volume}}$$

$$\frac{\frac{dy}{dx}\delta x}{\delta x} = \frac{dy}{dx}$$

The wave has thus created a volume strain in the air between the planes. Hence their is a variation of pressure from point to point along the tube.

Now, let p be the pressure-excess above normal on the plane *A* (now at *A'*). On *B* (now at *B'*) the pressure-excess will be $p + \frac{dp}{dx}\delta x$. These two pressure-excesses will act on the sample of air between *A'* and *B'* in the directions as shown. These are equivalent to:

(i) two equal and opposite pressure-excesses *p, p,* which provide the stress on the sample of air between *A'* and *B'* and produce the volume strain *dy/dx* in the sample, and

(ii) a resultant pressure-excess $(dp/dx)\,\delta x$ in the direction from *B'* to *A'*, which provides the restoring force causing the sample to come back to its initial position.

Let *K* be the bulk modulus of the air. Then by Hooke's law, we have

$$K = \frac{\text{stress}}{\text{strain}} = -\frac{p}{dy/dx}$$

negative sign showing that when pressure increases, the volume decreases.

$$\therefore \quad p = -K\frac{dy}{dx}$$

Now, the restoring force on the sample is $(dp/dx)\ \delta x$. Therefore, substituting for *p* from the last expression, we have

$$\text{restoring force} = \frac{dy}{dx}\left(-K\frac{dy}{dx}\right)\delta x = -K\,\frac{d^2y}{dx^2}K\,\delta x.$$

If ρ be the density of air, the mass of the sample will be $\rho\,\delta x$ (density x volume). Then, since force=mass x acceleration, the resultant force acting on the sample from B' to A' must be

$$\rho\,\delta x \times \left(-\frac{d^2y}{dt^2}\right)$$

where $-d^2y/dt^2$ is the instantaneous acceleration from B' *to* A'. Hence the equation of motion is

$$-K\frac{d^2y}{dx^2}\,\delta x = \rho\,\delta x \times \left(-\frac{d^2y}{dt^2}\right)$$

$$\frac{d^2y}{dx^2} = \frac{K}{\rho}\,\frac{d^2y}{dx^2}$$

Comparing it with the standard differential equation of wave motion$=\dfrac{d^2y}{dt^2} = v^2\dfrac{d^2y}{dx^2}$ we get

$$v^2 = \frac{K}{\rho}$$

$$v = \sqrt{\frac{K}{\rho}} \qquad \text{...(i)}$$

This gives the velocity of plane sound waves in a gas.

Newton's Formula: Newton assumed that the propagation of sound waves in a gas is an isothermal (constant temperature) process so that the Boyle's law holds. Thus, for a given mass of gas at pressure P and volume V, we have

$$PV = \text{constant.}$$

Differentiating :

$$PdV + VdP = 0$$

or

$$p = -\frac{dP}{dV/V}$$

But $-\dfrac{dP}{dV/V}$ represents the ratio $\dfrac{\text{volume stress}}{\text{volume strain}}$ which is the bulk modulus K of the gas.

$$\therefore \quad P = K$$

Substituting this value of K in eq. (i), we get

$$v = \sqrt{\frac{P}{\rho}}$$

This is Newton's formula. For air, at 0°C, we have

$$P = 76\times13.6\times981 \text{ dynes/cm}^2$$

and

$$\rho = 0.00129 \text{ gm/cm}^3.$$

$$v = \sqrt{\left(\frac{76\times13.6\times981}{0.00129}\right)} = 280 \text{ metre/sec}$$

which is much lower than the experimental value 332 metre/sec. Hence Newton's formula is not acceptable.

Laplace's Correction: Laplace pointed out that gases are bad conductors of heat, and the compressions and rarefactions occur very rapidly. Therefore the heat developed at compression cannot go to the surroundings, and the heat disappeared at rarefaction cannot come in from the surroundings. So the air in the path of a sound wave suffer temperature-changes, *i.e.* the conditions are *adiabatic* rather than isothermal. Thus the relation between pressure and volume of the gas through which a sound wave is passing is given by the adiabatic equation for a gas

$$PV^{\gamma} = \text{constant.}$$

Differentiating, $P\gamma\, dV^{\gamma-1}\, dV + V^{\gamma} - dP = 0$

or $\quad P\gamma\, dV + VdP = 0$

or $\quad \gamma P = -\dfrac{dP}{dV/V} = K.$

Substituting this value of K in eq. (i), we get

$$v = \sqrt{\frac{\gamma P}{\rho}} \quad \text{...(ii)}$$

This is Laplace's formula. Taking γ =1.41 for air; the velocity of sound in air comes out to be 331.6 metre/sec, which closely agrees with the experimental value.

Effect of Temperature

For 1 mole of a gas, we have

$$PV = RT$$

The mass of 1 mole of the gas is M gm, where M is molecular weight of the gas. Therefore, its density is given by

$$\rho = \frac{M}{V} = \frac{PM}{RT}$$

or
$$\frac{P}{\rho} = \frac{RT}{M}$$

Substituting this values of p/ρ in eq. (ii), we get

$$v = \sqrt{\frac{\gamma RT}{M}}$$

Thus, the velocity of sound is independent of pressure, but is proportional to the square-root of the absolute temperature of the gas. It is inversely proportional to the square-root of the molecular weight of the gas also.

If v_0 be the velocity at 0°C (=273 K.) and v_t the velocity at t°C {= (273 +t) K}, then

$$\frac{v_t}{v_o} = \sqrt{\left(\frac{273+t}{273}\right)} = \left(1+\frac{t}{273}\right)^{1/2}$$

or
$$\frac{v_t}{v_o} \simeq \left(1+\frac{1}{2}\frac{t}{273}\right) \simeq \left(1+\frac{t}{546}\right).$$

If $t = 1°C$, then

$$v_1 \simeq \left(1+\frac{1}{546}\right)$$

or

$$v_1 - v_0 \simeq \frac{v_0}{546}$$

For air, v_o=332 m/sec.

$$\therefore \qquad v_1 - v_0 \simeq \frac{332}{546} \simeq 0.61 \text{ m/sec.}$$

Thus the velocity of sound in air increases roughly by 0.61m/sec per degree centigrade rise in temperature.

Hebb's Telephone Method

An accurate method for measuring the velocity of sound in air is due to Hebb. It is based on an interference principle.

Two parabolic reflectors, R_1 and R_2 made of plaster of Paris, 5 feet in aperture and 15" in focal length are selected. They are placed coaxially face to face about 100 feet apart from each other. The reflector R is placed on a moving trolley so that it can be moved along its axis. At the focii F_1. F_2, of the reflectors are placed carbon microphones M_1, M_2. Each microphone is connected to a primary coil of a transformer through a battery, key and rheostat. The transformer has two primary coils P_1 and P_2, and a single secondary coil S to which is connected a telephone receiver T.

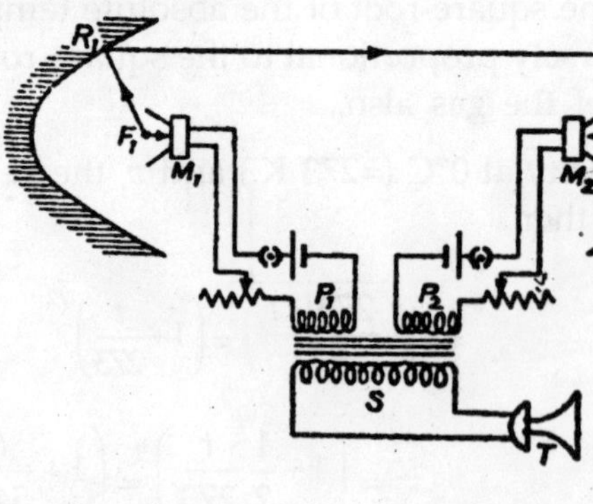

Working: A high-frequency whistle is blown at the focus ft. The microphone M_1 picks up the sound waves directly, while the microphone M_2 picks them up after they have reflected from R_x and then from R_2, as shown.

Thus, both M_1 and M_2 are set in vibration and convert the sound waves into electrical pulses with a phase difference depending on the path $F_1R_1R_2F_2$. These pulses pass through the primaries P_1 and P_2 and set up a resultant induced current in the secondary.

A resultant effect is therefore heard in the telephone *T*. When the path $F_1R_1R_2F_2$ is an even number of $\lambda/2$ (where λ is the wavelength), then the sounds falling on M_1 and M_2 will be in phase. Thus the two electrical currents in the secondary will be in phase and reinforce each other.

Hence a loud sound will be heard in *T*. When $F_1R_1R_2F_2$ is an odd number of $\lambda/2$, then the sounds, and hence the currents in the secondary, will be in opposite phase and the sound heard will be a minimum.

The reflector R_2 along with the microphone M_2 is moved along its axis. This causes a change in the path $F_1R_1R_2F_2$. Therefore, the sound heard in the Telephone *T* is alternately" maximum and minimum.

By properly adjusting the resistances in the two circuits, the minima may be made sharp. The position of the reflector R_2 at each minimum is noted, and the mean distance Between two consecutive minima is obtained. Clearly., this is λ. If *n* be the frequency of the whistle, the velocity of sound is obtained by $v = n\lambda$.

Problems

1. A metallic rod 2 metres long is firmly clamped in the middle and one end is stroked by a resined cloth. What is the pitch of the note which you will hear? Young's modulus = 20×10^{11} *dynes/cm*^{2} *and density=10 gm/cm*^{3}*.*

Solution: When the end of rod which is clamped in the middle is stroked, longitudinal stationary waves are set up in it with a node at the centre and antinodes at the ends. Hence the length l of the rod (distance between the antinodes) is $\lambda/2$.

$$\therefore \quad \lambda = 2l = 2\times200 = 400 \text{ cm.}$$

Now, the velocity of a longitudinal wave in a solid rod is

$$v = \sqrt{\frac{Y}{\rho}}$$

where Y is the Young's modulus and ρ the density of the material of the rod. Here 7=20×10[11] dynes/cm[2] and ρ =10 gm/cm[3].

$$\therefore \quad v = \sqrt{\frac{20\times10^{11}}{10}} = 4.472\times10^{5}\text{m/sec.}$$

If n be the frequency of the note emitted, we have

$$v = n\lambda$$

or $$n = \frac{v}{\lambda} = \frac{4.472\times10^{5}}{400} = 1118 \text{ cycle/sec.}$$

2. *At what temperature will the velocity of sound in air be doable its value at 0°C?*

Solution: Let t be the required temperature. Since the velocity is directly proportional to the square-root of the absolute temperature, we have

$$\frac{v_t}{v_o} = \sqrt{\frac{T}{T_0}} = \sqrt{\left(\frac{273+t}{273}\right)}$$

But $$v_1/v_0 = 2.$$

$$\therefore \quad 2 = \sqrt{\left(\frac{273+t}{273}\right)}$$

or $$4 = \frac{273+t}{273}$$

or $$t = (4\times273) - 273$$

$$= 819°\text{C.}$$

3. If the velocity of sound in hydrogen is 1300 metre/sec what will be the velocity in a mixture of 4 parts by volume of hydrogen and 1 part of oxygen at the same temperature and pressure ?

Solution: For any two gases the ratio of densities is equal to the ratio of their molecular weights. The density of mixture compared to hydrogen will be

$$\rho = \frac{\text{mass}}{\text{Volume}} = \frac{V_h\rho_h + V_o\rho_o}{V_h + V_o}$$

$$= \frac{(4V \times 1) + (V \times 16)}{4V + V} = 4.$$

The velocity is inversely proportional to the square-root of density.

$$\therefore \qquad \frac{v}{v_h} = \sqrt{\frac{\rho_h}{\rho}}$$

$$v = 2_h\sqrt{\frac{\rho_h}{\rho}}$$

$$= 1300 \times \sqrt{\frac{1}{4}} = 650 \text{ m/sec.}$$

4. The mean molecular weight of air is 28.8. If γ is 1.4, calculate the velocity of sound in air at 27°C. Gas constant R is 8.31×10^7erg/ (mole – K).

Solution: The velocity of sound in a gas, in terms of the gas molecular weight, is given by

$$v = \sqrt{\frac{\gamma RT}{M}}$$

Substituting the given values, we get

$$v = \sqrt{\frac{1.4 \times (8.31 \times 10^7 \, erg/mole - K) \times 300K}{28.8gm}}$$

$$= 3.48 \times 10^4 \text{ cm/sec}$$

$$= 348 \text{ metre/sec.}$$

5. If the velocity of sound in hydrogen be 1300 m/sec at a certain temperature, what will be the velocity at the same temperature in a diatomic gas of molecular weight 32 ?

Solution: Hydrogen is a diatomic gas, and γ is same for all diatomic gases. Now,

$$v = \sqrt{\frac{\gamma RT}{M}}$$

so that at constant temperature,

$$v \propto \frac{1}{\sqrt{M}}$$

The molecular weight of hydrogen gas (H_2) is 2. Thus, if v be the required velocity, we have

$$\frac{v}{1300} = \sqrt{\frac{2}{32}}$$

$$v = 1300 \times \frac{1}{4} = 325 \text{ metre/sec.}$$

9

Stationary Waves

Standing Waves

When two identical waves, either transverse or longitudinal, travel through a medium along the same line in opposite directions, they superpose to produce a new type of waves which appear stationary in space. These waves are called "stationary waves" or "standing waves". For example, when a wave is sent along a string or along the air-column of a pipe, it is reflected at the other end and superimpose upon the incident wave to produce stationary waves. When such waves are formed, then certain particles of the medium remain permanently at rest, while some other particles undergo maximum displacement compared to others. The former are called the "nodes" and the later the "antinodes".

Conditions for Stationary Waves: Obviously, for the formation of stationary waves, the medium must not be infinite in length ; it should have a boundary. In such a medium, any travelling wave will, by reflection at the boundary, produce an identical wave travelling in the opposite direction so that stationary waves will be formed. Hence the presence of a "bounded" medium is the essential condition for the formation of stationary waves.

Formation of Stationary Waves and their Characteristic Properties

Theoretical Investigation

Let us consider a plane progressive wave of amplitude a travelling with velocity v along the $+x$ direction. Its equation may be written as

$$y_1 = a \sin \frac{2\pi}{\lambda} (vt - x),$$

where y_1 is the displacement at a point x at an instant t. The length of the wave is λ.

Let this wave be reflected from a rigid boundary wall situated at x=0. The displacement at the point x at an instant t due to the reflected wave moving along the $-x$ direction will be given by

$$y_2 = a' \sin \frac{2\pi}{\lambda} (vt+x),$$

where a' is the reflected amplitude. By the principle of superposition, the resultant displacement at the point x at the instant t would be given by

$$y = y_1 + y_2 = a \sin \frac{2\pi}{\lambda} (vt-x) + a' \sin \frac{2\pi}{\lambda} (vt-x).$$

The displacement (y) at the "rigid" "wall (at x=0) will always be zero. Therefore, putting y=0 and $x = 0$ in the last equation, we have

$$O = a \sin \frac{2\pi}{\lambda} (vt) + a' \sin \frac{2\pi}{\lambda} (vt)$$

or
$$a' = - a.$$

Hence the resultant displacement is

$$Y = a\left[\sin\frac{2\pi}{\lambda}(vt - x) - \sin\frac{2\pi}{\lambda}(vt + x)\right]$$

$$= a\left[2\cos\frac{2\pi vt}{\lambda}\sin\left(-\frac{2\pi x}{\lambda}\right)\right]$$

$$= -\,2\,a\,\sin\frac{2\pi x}{\lambda}\cos\frac{2\pi vt}{\lambda}$$

This represents the resultant disturbance which is of the same period as the original waves but of amplitude $-2a$ sin $(2\pi x/\lambda)$. This is the equation of the resultant stationary wave.

The resultant particle velocity is given by

$$\frac{\partial y}{\partial t} = \frac{4\pi av}{\lambda}\sin\frac{2\pi x}{\lambda}\sin\frac{2\pi vt}{\lambda} \quad \text{...(ii)}$$

The resultant pressure-variation is given by

$$\rho = -\,K\,\frac{\partial y}{\partial t}$$

$$= K = \frac{4\pi a}{\lambda}\cos\frac{2\pi x}{\lambda}\cos\frac{2\pi vt}{\lambda} \quad \text{...(iii)}$$

Eq. (i), (ii) and (iii) show that the resultant displacement, velocity and pressure-variation in the stationary wave change with position *(x)* as well as with time *(t)*.

Change with Position

At points where $\frac{2\pi x}{\lambda}$ = s0, π ,2-,...

$$x = 0, \frac{\lambda}{2}, \frac{2\lambda}{2} \quad \text{...(iv)}$$

we have sin $(2\pi x/\lambda) = 0$ and cos $(2\pi x/\lambda) = \pm 1$. Therefore eq. (i), (ii) and (iii) show that at these points the displacement y and the velocity *dy/dt* are always zero, while the pressure variation ρ is always maximum (alternately positive and negative) relative to all other points. These points are the "nodes" and are separated from each other by $\lambda/2$, as shown by eq. (iv). At the fixed wall $x = 0$ there is a node.

At points where $\frac{2\pi x}{\lambda} = \frac{\pi}{2}, \frac{3\pi}{2}, \frac{5\pi}{2}, ...$

$$x = \frac{\lambda}{4}, \frac{3\lambda}{4}, \frac{5\lambda}{4}, ... \qquad ...(v)$$

we have $\sin(2\pi x/\lambda)=\pm 1$ and $\cos(2\pi x/\lambda)=0$. Again eq. (i), (ii), and (iii) show that at these points the displacement y and the velocity $\partial y/\partial t$ are always maximum, while the pressure-variation p is always minimum (zero) relative to all other points. These points are the "antinodes," and are separated from each other by $\lambda/2$, as shown by eq. (v).

Change with Time

At instants when $\frac{2\pi vt}{\lambda} = \frac{2\pi t}{T} = 0, \pi, 2\pi, ...$ $\qquad [\because \lambda = vT]$

$$t = 0, \frac{T}{2}, \frac{2T}{2}, ...$$

we have $\cos(2\pi vt/\lambda)=0$ and $\sin(2\pi vt/\lambda) = 0$. eq. (i), (iii) and (ii) show that at these instants the displacement y and the pressure-variation p are everywhere maximum, while the velocity $\partial y/\partial t$ is everywhere minimum (zero) relative to all other instants. Such instants occur twice in each period.

At instants when $\frac{2\pi vt}{\lambda} = \frac{2\pi t}{T} = \frac{\pi}{2}, \frac{3\pi}{2}, \frac{5\pi}{2}, ...$

$$t = \frac{T}{4}, \frac{3T}{4}, \frac{5T}{4}, ...$$

we have $\cos(2\pi vt/\lambda)=0$ and $\sin(2\pi vt/\lambda)=\pm 1$. Again eq. (i), (iii) and (ii) show that at these instants the displacement y and the pressure-variation p are everywhere zero, while the velocity $\partial y/\partial t$ is everywhere maximum relative to all other instants. Such instants also occur twice in each period.

Thus the superposition of incident and reflected waves produce stationary waves characterised by nodes which remain permanently at rest but suffer maximum pressure-variation, and the antinodes which always have maximum displacement but suffer minimum pressure variation.

Energy Transmission in Stationary Waves: We know that in stationary waves the pressure-variation p in the medium is given by

$$p = K\frac{4\pi a}{\lambda}\cos\frac{2\pi x}{\lambda}\cos\frac{2\pi vt}{\lambda}$$

$$= p_x \cos\frac{2\pi vt}{\lambda},$$

where $$p_x = k\frac{4\pi a}{\lambda}\cos\frac{2\pi x}{\lambda}$$

Similarly, the particle velocity u is given by

$$u = \frac{\partial y}{\partial t} = \frac{4\pi av}{\lambda}\sin\frac{2\pi x}{\lambda}\sin\frac{2\pi vt}{\lambda}$$

$$= u_x \sin\frac{2\pi vt}{\lambda},$$

where $$u_x = \frac{4\pi av}{\lambda}\sin\frac{2\pi x}{\lambda}.$$

Therefore, the energy transmitted (work done) per unit area in a small interval of time dt

$$= pudt.$$

Hence, the energy transmitted in the periodic time T

$$= \int_0^T pudt.$$

$\therefore$ rate of energy transmission

$$= \frac{1}{T}\int_0^T pudt.$$

$$= \frac{1}{T}\int_0^T p_x \cos(2\pi vt/\lambda)\, u_x \sin(2\pi vt/\lambda)\, dt$$

$$= \frac{P_x U_x}{T}\int_0^T \sin(2\pi vt/\lambda)\cos(2\pi vt/\lambda)\, dt$$

$$= \frac{P_x U_x}{2T}\int_0^T \sin(4\pi vt/\lambda)\, dt$$

$$= \frac{P_x U_x}{2T}\int_0^T \sin\frac{4\pi t}{T}\, dt \qquad \left[\because v = \frac{\lambda}{T}\right]$$

$$= \frac{P_x U_x}{2T}\left[-\frac{T}{4\pi}\cos\frac{4\pi t}{T}\right]_0^T \left[-\frac{T}{4\pi}\cos\frac{4\pi t}{T}\right]_0^T$$

$$= \frac{P_x U_x}{8\pi}[1-1]$$

$$= 0.$$

Thus in stationary waves there is no net transmission of energy across any plane in either direction. The energy remains standing in the medium, although it alternates between vibrational kinetic energy and elastic potential energy.

Characteristics of Stationary Waves

The above theory tells us the following characteristics in stationary waves :

(i) All particles of the medium oscillate at the same time and with the same frequency.

(ii) The amplitude of oscillation varies from zero at the nodes to a maximum at the antinodes.

(iii) Consecutive nodes, us well as consecutive antinodes, are $\lambda/2$ apart. Nodes and antinodes occur alternately.

(iv) The nodes remain permanently at rest but (in case of longitudinal stationary waves) suffer maximum pressure-variation and are alternately the centre of compression and the centre of rarefaction.

(v) The antinodes have always maximum displacement and maximum velocity relative to other points but (in case of longitudinal stationary waves) do not suffer pressure-variation.

(vi) All the points between any pair of nodes are in same phase, so that they all reach their maximum displacements at the same instant, but the phase suddenly reverses at each node.

(vii) Twice in one period all the particles simultaneously pass through their mean positions. The direction of motion is reversed after each half-period.

(viii) Twice in one period all the particles simultaneously have their maximum displacements and are momentarily at rest. The displacements are reversed after each half-period.

(ix) There is no net flow of energy in any direction.

Partial Reflection: If the incident travelling wave is only partially reflected from the boundary, the amplitude of the reflected wave will be smaller than that of the incident wave because some of the incident energy has been transmitted into the second medium. The superposition of two waves of unequal amplitudes and travelling in opposite directions would give a standing-wave pattern, but the nodes will have a finite amplitude, and there will be a net flow of energy (though small) from the first medium into the second.

Distinction between Progressive and Stationary Waves

Progressive Waves	*Stationary Waves*
1. These are advancing waves travelling in the medium with a finite velocity. The disturbance travels onwords, being handed over from point to point.	1. There is no advancement of the waves in either direction.
2 All points of the medium vibrate with the same amplitude.	2. The amplitude varies from zero at a node to maximum at the antinode.
3. The phase of vibration varies continuously from point to point. The instantaneous arrangement of the points repeats each after a distance λ.	3. All points between any two consecutive nodes vibrate in the same phase, but the phase suddenly reverses at each node.

Contd...

Progressive Waves	*Stationary Waves*
4. No point is permanently at rest. Each point is how ever, momentarily at rest at its maximum displacement.	4. The nodes are permanently at rest. Other points are momentarily at rest at their maximum displacement.
5. Different points reach the position of maximum displacement at different times.	5. All points reach the position of maximum displacement simultaneously twice in each period.
6. All points cross their mean positions in succession but with the same velocity.	6. All points cross their mean positions simultaneously but with different velocities. The velocity is maximum at the antinodes and zero at the nodes
7. The pressure-variation, being same at all points of the medium, travels forward.	7. The pressure-variation is different at different points. It is maximum at the nodes and minimum at the antinodes. It does not move forward but occurs simultaneously at all places.
8. Energy flows across every plane in the direction of propagation of the wave.	8. There is no flow of energy across any plane.

Waves in a Linear Bounded Medium

A linear medium is one in which a wave travels along a fixed line. Transverse waves travelling on a stretched string, longitudinal waves in a solid rod and sound waves in a column of air in a pipe are the examples of waves in a linear medium. Now, any one-dimensional wave must satisfy the general differential equation

$$\frac{\partial^2 y}{\partial t^2} = v^2 \frac{\partial^2 y}{\partial x^2}$$

where y is either a transverse or a longitudinal displacement. The solution of this equation will, of course, be a function of the variables x and t. A general solution may be written as

$$y = f_1\,(vt - x) + f_2\,(vt + x). \qquad \text{...(ii)}$$

It simply means that the resultant displacement at any

point is in general due to two waves, one travelling along the $+x$ direction and the other along the $-x$ direction.

In case of a linear medium of infinite length, the functions f_1 and f_2 depend only on the initial disturbance responsible for creating the wave. If the medium is not infinite, it is said to be bounded. Then f_1 and f_2 are governed by the boundary conditions.

Discrete Frequencies of a Linear Bounded Medium: Let us consider a linear medium rigidly bounded at $x=0$ and at $x=l$. Let us consider the simplest case of a harmonic wave of amplitude a moving with velocity v in the $+x$ direction and an identical (reflected) wave of amplitude a' moving in the $-x$ direction. Then the solution of eq. (i) will be of the form

$$y = a \sin\frac{2\pi}{\lambda}(vt - x) + a' \sin\frac{2\pi}{\lambda}(vt+x), \qquad \text{...(iii)}$$

with the boundary condition that $y=0$ at $x=0$ and also at $x=l$ at all times.

The condition $y = 0$ at $x = 0$ gives for all values of t,

$$0 = a \sin\frac{2\pi vt}{\lambda} + a' \sin\frac{2\pi vt}{\lambda}$$

so that $a' = -a$

Thus
$$y = a\left[\sin\frac{2\pi}{\lambda}(vt-x) - \sin\frac{2\pi}{\lambda}(vt+x)\right]$$

$$= a\left[2\cos\frac{2\pi vt}{\lambda}\sin\left(-\frac{2\pi x}{\lambda}\right)\right]$$

$$= -2a \sin\frac{2\pi x}{\lambda}\cos\frac{2\pi vt}{\lambda} \qquad \text{...(iv)}$$

This expression represents stationary waves.

Let us now apply the other condition $y=0$ at $x=l$ for all values of t. This gives

$$0 = -2a \sin\frac{2\pi l}{\lambda}\cos\frac{2\pi vt}{\lambda}.$$

Since this must hold for all values of t; it must be that

$$\sin\frac{2\pi l}{\lambda} = 0$$

or $$\frac{2\pi l}{\lambda} = p\,\pi, \qquad p = 0,1,2,\ldots\ldots$$

or $$\lambda = \frac{2l}{p},$$

limiting the values of allowed frequencies to

$$n = \frac{v}{\lambda} = \frac{pv}{2l}.$$

Substituting the above value of A in eq. (iv), we get

$$y = -\,2a\,\sin\frac{p\pi x}{l}\,\cos\frac{p\pi vt}{l}$$

This is the solution of the differential equation of wave motion for a medium bounded at both ends. This shows that in such a medium only vibrations of certain discrete frequencies are possible $\left(n = \frac{v}{2l}, \frac{2v}{2l}, \frac{3v}{2l}, \ldots\right)$. These frequencies are the "normal frequencies" or "modes of vibration". The lowest frequency is the "fundamental frequency". Since the frequencies are in a harmonic series, they are also called "harmonics". We see that in a linear bounded medium when both the boundaries are rigid, all the harmonics even and odd are present. The same is true when both the boundaries are free.

One Boundary Rigid and the other Free: Let us consider a linear bounded medium having one boundary rigid at $x=0$ and the other boundary free at $x=l$. (An example of such a medium is a pipe with one end closed and the other open, or a solid rod clamped at one end). The boundary conditions are

$$y=0 \text{ at } x=0 \text{ and } \frac{\partial y}{\partial x}=0 \text{ at } x=l \text{ for all values of } t.$$

Now, the displacement at any point x at time t may be written a

$$y = a\,\sin\frac{2\pi}{\lambda}(vt-x) + a'\,\sin\frac{2\pi}{\lambda}(vt+x).$$

The condition $y = 0$ at $x = 0$ for all values of t gives

$$a' = -a,$$

so that

$$s = a\left[2\cos\frac{2\pi vt}{\lambda}\sin(-\frac{2\pi x}{\lambda})\right]$$

$$= -2a\sin\frac{2\pi x}{\lambda}\cos\frac{2\pi vt}{\lambda} \quad \text{...(i)}$$

Differentiating partially with respect to x, we get

$$\frac{\delta y}{\delta x} = -\frac{4\pi a}{\lambda}\cos\frac{2\pi x}{\lambda}\cos\frac{2\pi vt}{\lambda}.$$

Applying the condition

$$\frac{\partial y}{\partial x} = 0 \text{ at } x=l:$$

$$0 = -\frac{4\pi a}{\lambda}\cos\frac{2\pi l}{\lambda}\cos\frac{2\pi vl}{\lambda}.$$

Since this must hold for all values of t; it must be that

$$\cos\frac{2\pi l}{\lambda} = 0$$

or

$$\frac{2\pi l}{\lambda} = (2\rho+1)\frac{\pi}{2} \qquad p = 0, 1, 2, \ldots.$$

or

$$\lambda = \frac{4l}{2p=1}$$

The allowed frequencies are thus

$$n = \frac{v}{\lambda} = \frac{(2p+1)v}{4l}$$

$$= \frac{v}{4l}, \frac{3v}{4l}, \frac{5v}{4l} \ldots\ldots$$

The fundamental frequency is $v/4l$, and the higher frequencies are odd multiples of it. In other words, the frequencies of the harmonics are proportional to the odd numbers 1, 3, 5,....Thus only odd harmonics are present when one boundary is rigid and the other is free.

Both Boundaries Free: Let us now consider a medium with free boundaries at $x=0$ and $x=l$. (An example of such a medium is a pipe open at both ends, or a solid rod clamped at the middle.) The boundary conditions are

$$\frac{\partial y}{\partial t}=0 \text{ at } x=0 \text{ and also at } x=l \text{ for all values of } t.$$

Let us consider the displacement at any point x at time t:

$$y = a \sin\frac{2\pi}{\lambda}(vt-x)+a' \sin\frac{2\pi}{\lambda}(vt+x).$$

Differentiating partially with respect to x:

$$\frac{\partial y}{\partial x} = -\frac{2\pi a}{\lambda}\cos\frac{2\pi}{\lambda}(vt-x)+\frac{2\pi a'}{\lambda}\cos\frac{2\pi}{\lambda}(vt+x)$$

Applying the condition $\frac{\partial y}{\partial x}=0$ at $x=0$, we have

$$0 = -\frac{2\pi a}{\lambda}\cos\frac{2\pi vt}{\lambda}+\frac{2\pi a'}{\lambda}\cos\frac{2\pi vt}{\lambda}$$

This gives $\qquad a' = a.$

Thus
$$y = a\left[\sin\frac{2\pi}{\lambda}(vt-x)+\sin\frac{2\pi}{\lambda}(vt+x)\right]$$

$$= a\left[2\sin\frac{2\pi vt}{\lambda}\cos(-\frac{2\pi x}{\lambda})\right]$$

$$= 2a\cos\frac{2\pi x}{\lambda}\sin\frac{2\pi vt}{\lambda}. \qquad \text{...(ii)}$$

Again, differentiating partially:

$$\frac{\partial y}{\partial x} = -\frac{4\pi a}{\lambda}\sin\frac{2\pi x}{\lambda}\sin\frac{2\pi vt}{\lambda}$$

Applying the condition $\frac{\partial y}{\partial x}=0$ at $x=l$, we get

$$0 = -\frac{4\pi a}{\lambda}\sin\frac{2\pi l}{\lambda}\sin\frac{2\pi vt}{\lambda}$$

Since this must hold for all values of t, it must be that

$$\sin\frac{2\pi l}{\lambda} = o$$

$$\frac{2\pi l}{\lambda} = p\pi \qquad p=0,1,2.....$$

$$\lambda = \frac{2\pi t}{\lambda}$$

The allowed frequencies are

$$n = \frac{v}{\lambda} = \frac{pv}{2l}$$

$$= \frac{v}{2l}, \frac{2v}{2l}, \frac{3v}{2l}, \ldots\ldots$$

The fundamental frequency is $v/2l$, and the higher frequencies are integral multiples of it. Thus all the harmonics arc present in this case.

Problems

1. *A string vibrates according to the equation*

$$y = 5 \sin\frac{\pi x}{3} \cos 40\pi t,$$

where x and y are in cm and t in second. Find oat the amplitude and velocity of the two component waves whose superposition can give rise to this vibration. What is the distance between adjacent nodes ? What is the velocity of a particle of the string at x=l.5 cm at t=9/8 second.

Solution: The given equation

$$y = 5 \sin\frac{\pi x}{3} \cos 40\pi t, \qquad \text{...(i)}$$

represents a stationary wave. It may be written as

$$y = \frac{5}{2}\sin\left(40\pi t + \frac{\pi x}{3}\right)$$

$$-\frac{5}{2}\sin\left(40\pi t-\frac{\pi x}{3}\right)$$

The two sine terms represent two harmonic waves travelling in opposite directions. These two waves constitute the given stationary wave. The component waves may be written as

$$y_l = \frac{5}{2}\sin\frac{2\pi}{6}(120t + x)$$

and $$y_l = -\frac{5}{2}\sin\frac{2\pi}{6}(120t - x).$$

Comparing the component wave equations with the standard equation, $y = \pm\, a \sin \frac{2\pi}{\lambda}(vt \pm x)$, we see that

amplitude, $a = \frac{5}{2}$ cm,

wavelength, $\lambda = 6$ cm,

velocity, $v = 120$ cm/sec,

and frequency, $n = \frac{v}{\lambda} = 20\ \text{sec}^{-1}$.

The distance between adjacent nodes is

$$\frac{\lambda}{2} = 3 \text{ cm.}$$

Differentiating eq. (i) with respect to t, we get

$$\frac{\partial y}{\partial t} = -5\times 40\,\pi \sin\frac{\pi x}{3}\sin 40\,\pi\, t.$$

At x=1.5 cm and $t=\frac{9}{8}$ sec, we have

$$\frac{\partial y}{\partial t} = -\,200\,\pi \sin\frac{\pi}{3}\sin 45\,\pi$$

$$= 0. \qquad [\because \sin 45\,\pi = 0]$$

2. *The equation of a stationary wave is*

$$x = 6\cos\left(\frac{\pi x}{3}\right)\sin 60\pi t \text{ cm.}$$

Find the (i) equation of waves constituting the stationary wave (ii) wavelength and frequency of these waves, (iii) distance between successive nodes.

Hint : $\text{Sin } A + \sin B = 2\sin\frac{A+B}{2}\cos\frac{A+B}{2}$.

[Ans. (i) $x_1 = 3\sin\left(60\pi t + \frac{\pi x}{3}\right)$ $x_2 = 3\sin\left(60\pi t - \frac{\pi x}{3}\right)$; (ii) 6 cm, 30 sec^{-1}, (iii) 3 cm.]

3. *A longitudinal stationary wave in a rod is expressed as*

$$y = 0.002 \sin (0.15\, x) \sin (4.8\text{x}10^4\, t).$$

Calculate the maximum particle velocity and maximum tensile stress at the point x=3.5 cm. The Young's modulus of the material of the rod is 8.0×10^{11} dyne/cm^2.

Solution: The particle velocity at position x and time t is obtained by differentiating the given equation with respect to t:

$$\frac{\partial y}{\partial t} = 4.8\times10^4\,(0.002)\sin(0.15x)\cos(4.8\times10^4\,t).$$

The maximum velocity at x occurs when $\cos(4.8\times10^4\, t)=1$. Thus

$$\left(\frac{\partial y}{\partial t}\right)_{max} = 4.8\times10^4\,(0.002)\sin(0.15x)$$

$$= 96\sin(0.15\,x).$$

At x=3.5 cm, we have

$$0.15x = 0.15\times3.5 = 0.525 \text{ radian}$$

$$= 0.525\times\frac{180^\circ}{3.14} \qquad [\because \pi \text{ rad}=180^\circ]$$

$$= 30^\circ$$

$$\therefore \left(\frac{\partial y}{\partial t}\right)_{max} = 96 \sin (30°) = 48 \text{ cm/sec.}$$

In longitudinal waves the tensile strain is given by $\partial y/\partial x$. So the dy tensile stress is Young's modulus×tensile strain, that is, $Y\frac{\partial y}{\partial x}$. Now, differentiating the given equation with respect to x, we get

$$\frac{\partial y}{\partial x} = 0.15\ (0.002) \cos (0.15\ x) \sin (4.8\times10^4\ t).$$

$$\therefore \text{ tensile stress} = Y\frac{\partial y}{\partial x}$$

$$= (8.0\times10^{11} \text{ dyne/cm}^2)\ (3\times10^{-4}) \cos (0.15\ x) \sin (4.8\times10^4\ t).$$

The tensile stress at x is maximum when $\sin (4.8\times10^4\ t)=$
Thus

$$\text{max. tensile stress} = (8.0\times10^{11} \text{ dyne/cm}^2)\ (3\times10^{-4}) \cos (0.15\ x)$$
$$= 24\times10^7 \cos (0.15\ x) \text{ dyne/cm}^2.$$

At x=3·5 cm, we have

$$\text{max. tensile stress} = 24\times10^7 \cos 30°$$
$$= 24\times10^7 \text{X } 0.866$$
$$= 2.1\times10^8 \text{ dyne/cm.}$$

4. A stationary wave produced in a rod is expressed as

$$y = 0.01 \sin (0.13\ x) \sin (200\ t) \text{ cm.}$$

Calculate the maximum particle velocity and tensile stress at the point x=4 cm. The Young's modulus of the material of the rod is 10^{12} dyne/cm².

[Ans. 1.0 cm/sec, 1.1×10^9 dyne/cm²].

10

Wave Motion

Pressure and Energy

A Mechanical Progressive Wave: "A mechanical progressive wave is a form of disturbance which travels from one part of an elastic material medium to the other through the oscillatory motion of the particles of the medium about their equilibrium positions." Water waves, sound waves, waves in a spring etc. are the examples of mechanical waves.

Chief Characteristics: Let us consider a number of particles forming a continuous *elastic* material medium. If the first particle is disturbed and set in oscillation, the disturbance is handed over from particle to particle due to the elasticity of the medium. That is, each particle is set in oscillation but, due to *inertia,* a little later than the previous particle. Thus *the phase of oscillation changes from particle to particle.* When the disturbance reaches a particle at the moment when the first particle has completed one oscillation, then this particle oscillates in the same phase as the first particle.

Thus, in a mechanical wave some kind of disturbance, or energy or momentum, is transmitted from one place to another through a material medium, but the medium itself is not transported. In fact, it is the phase relationship of the medium particles we observe as waves. All material media-solids, liquids and gases-can carry energy by means of waves.

We can demonstrate the wave characteristics by dropping a stone into a quiet lake in which there is a floating piece of wood. Circular waves spread out from the point of impact and travel over the surface of the lake. When they reach the piece of wood, they set it in up and down motion. This shows that the waves have transferred energy and momentum to the piece of wood. The destructive power of an ocean tide is also a good demonstration of the energy transported by water waves.

Transverse and Longitudinal Waves

There are two types of waves :

Transverse Wave: *A transverse wave is one in which the direction of oscillation of the particles of the medium is at right angles to the direction of propagation of the wave.* For example, the waves generated in ropes and stretched strings and the light (electromagnetic) waves are transverse waves.

Longitudinal Wave: *A longitudinal wave is one in which the direction of oscillation of the particles of the medium is the same as the direction of propagation of the wave.* For example, if a spring fixed at one end is pulled at the other and then let free, a longitudinal wave is generated in it. Sound waves are also longitudinal waves.

Formation of Transverse Waves: Let us consider a number of equidistant particles *a*, *b*, c,... of an elastic medium along a straight line. Suppose at time $t=0$ the particle *a* is disturbed so as to perform a simple harmonic oscillation at right angles to this line. Let T be the period of oscillation. As the particle *a* moves up, the particles *b*, *c* are dragged on, one after the other. Suppose at $t=T/4$ when the particle a has attained its maximum upward displacement, the disturbance reaches the particle *c* which is now just to start. At *t=T/2,* the particle *a* comes to its equilibrium position, *c* attains its maximum upward displacement and the particle *e* is just disturbed. At *t=3T/4,* the particle a attains its maximum downward displacement, *c* comes back to its equilibrium position, *e* attains its maximum upward displacement and *g* is just disturbed. At *t=T,* the particle *a* once again comes to its equilibrium position completing one full oscillation, and the disturbance reaches the particle *i*.

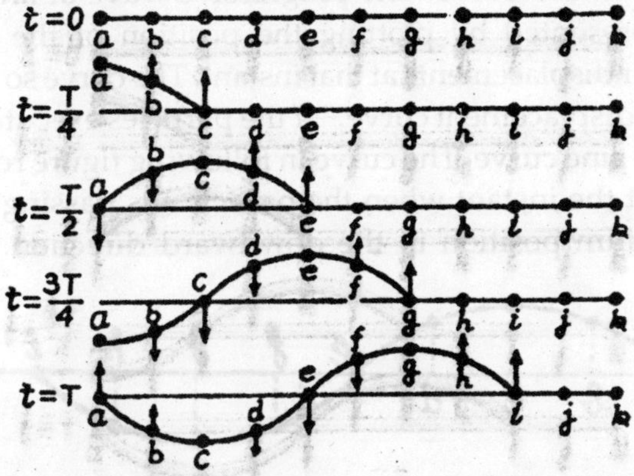

The particle *i* now oscillates in the same phase as the particle *a*. As the particle *a* performs its oscillation again and again, a wave-train is propagated in the medium.

The 'elasticity' and the 'inertia' of the medium play important parts in the propagation of wave. The elasticity brings the particles momentarily at rest in their extreme positions, while inertia carries them across their equilibrium positions.

The maximum displacement of each particle in either direction is called the "amplitude" of the wave.

The distance between two nearest particles oscillating in the same phase is called the "wavelength" (λ). The number of oscillations made in one second by a particle is called the "frequency" (*n*). As a particle completes one oscillation, the wave covers a distance λ. Hence the velocity of propagation of the wave is given by

$$v = \frac{\text{distance}}{\text{time}} = \frac{\lambda}{T}$$

But $\frac{1}{T} = n$, the "frequency" of oscillation.

$$\therefore \qquad v = n\lambda$$

Displacement Curve: The progressive wave at any instant can be represented by plotting the position of the particles against their displacements at that instant. The curve so obtained is called a "displacement curve." If the particles execute S.H.M., we obtain a sine curve. The curve in following figure represents the wave at the instant when the particle *a* is passing through its equilibrium position in the downward direction.

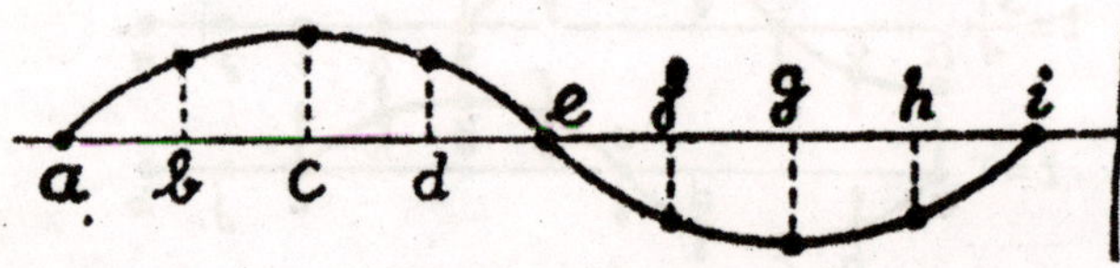

Formation of Longitudinal Waves: Compressions and Rarefactions: Let us consider the propagation of a longitudinal wave in air. The dots *a, b, c, ...*represent air particles. When a longitudinal wave passes, say from left to right, each particle turn by turn begins to oscillate along the direction of propagation. Let *T* be the period of oscillation. Suppose the disturbance takes *T/8* second in reaching from one particle to the next.

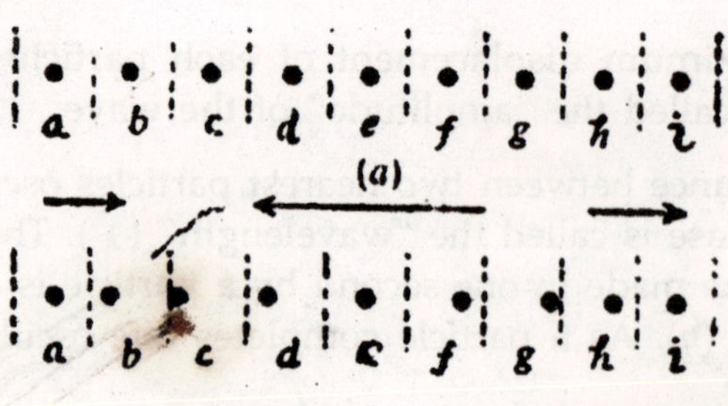

Then the motion of any particle will lag behind the motion of its left-hand neighbour by T/8 second. Thus when the particle *a* is passing through its equilibrium position from left to right, the particle *b* will also be moving from left to right but will reach its equilibrium position T/8 second later. The particle *c* will be *T*/8 second behind *b* (or T/4 second behind a) and hence it will be momentarily at rest at its extreme left position.

The particle *d* will be moving towards its extreme left position, and so on. Thus we shall find that at the instant considered the particles *a, b, h,* and *i* will be moving towards right; *d, e* and/towards the left; and *c* and *g* are momentarily at rest. We know that each particle is lagging behind its left-hand neighbour by *T*/8 second. The separation between different pairs of neighbours is, however, different because the velocity of different particles at any instant is different. Those particles which are passing through their equilibrium positions have greatest velocity while those at the end of their swing are momentarily at rest.

The figure shows that at the instant considered the particles from *a* to c are closer than when undisturbed. Thus they form a region of 'compression' which is greatest at *a*. The particles from *c* to *g* are separated more than when undisturbed and form a region of 'rarefaction' which is greatest at *e*. At *g* the rarefaction again changes to compression, and so on. Thus we can say that a longitudinal wave in air consists of alternate compressions and rarefactions.

Displacement Carve: In a longitudinal wave the direction of vibration of the particles of the medium is the same as the direction of propagation of the wave.

If, at any instant, we represent the forward displacements of the particles along a perpendicular direction above their equilibrium positions, and the backward displacements below the equilibrium positions, we obtain a displacement curve having the same shape as a sine curve. This curve may be taken to represent the longitudinal wave at the particular instant.

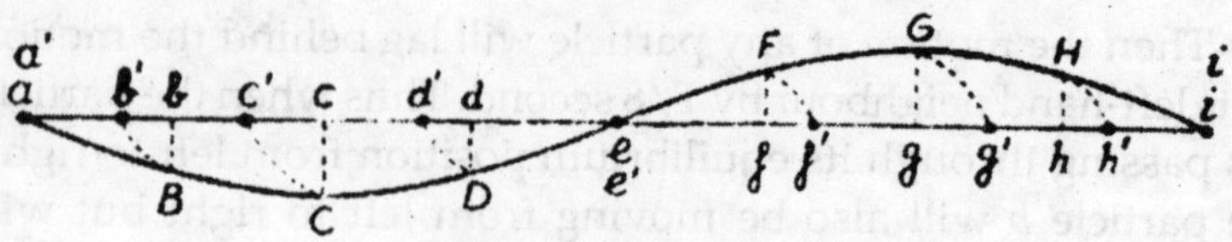

Direction of Wave

Let, in figure above *a, b, c,* be the equilibrium positions of medium particles, and *a',b',c'*,their displaced positions at any instant during the passage of the wave. Particles *a, e* and *i* are at their equilibrium positions at this instant. Particles *b, c, d* are displaced backwards while particles *f, g, h* are displaced forwards.

Let us draw perpendiculars *bB(=bb')* ; *cC(=cc')* ; *dD(=dd')* in the downward direction and perpendiculars *fF(=ff'); gG(=gg'); hH(==hh')* in the upward direction. On joining *aBCDeFGHi,* we obtain a sine curve. This is the "displacement curve" for the longitudinal wave.

Propagation of Sound Waves

When a sound source vibrates, it generates waves in the surrounding air. To understand the phenomenon, let us consider a sounding tuning fork. As the prongs of the fork move outward, they push away the air in their contact and cause compression.

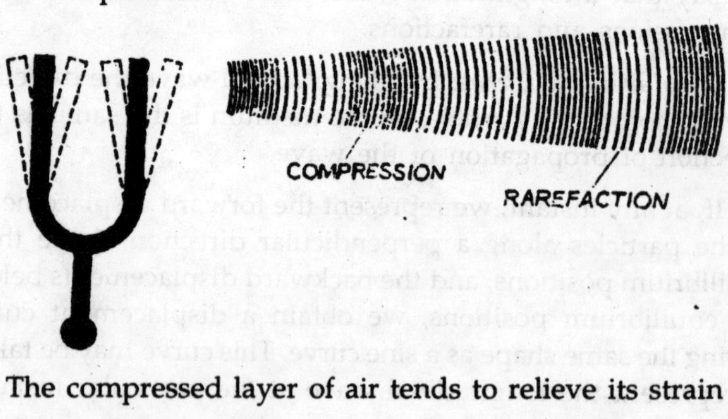

The compressed layer of air tends to relieve its strain by

moving away from the compression, thus compressing the next layer. This layer does likewise and pass on the compression to the next layer, and so on.

When the prongs move inward, the layer in their immediate contact is rarefied and there is a local fall in pressure. The layer to the right of the rarefaction moves in to make up the loss in pressure and so the rarefaction is also passed from layer to layer.

The phenomenon is repeated with each vibration of the fork. Thus a train of waves consisting of alternate compressions and rarefactions, that is, fluctuations of air density, at regular intervals travels onward in the air. We know that the occurrence of compressions and rarefactions in the medium is the characteristic of longitudinal waves. Hence sound waves are longitudinal.

Sound Waves and Light Waves: Sound waves are mechanical waves which are transmitted in some material medium. In contrast to them there art light waves which are electro-magnetic waves in which no material medium is involved.

These waves are radiated by an oscillating electric charge. The oscillating charge produces continually-changing electric and magnetic fields in its immediate vicinity. These changing fields produce other fields in their vicinity and the disturbance travels outward with the speed of light. Light waves are transverse waves.

General Equation of Progressive Waves

Let us consider a series of particles forming a long string stretched along the X-axis. Suppose the particle at the origin *O* acts as the source of a disturbance which travels as a transverse wave in the positive direction of the X-axis with a velocity v. As the wave proceeds, each successive particle is set in motion.

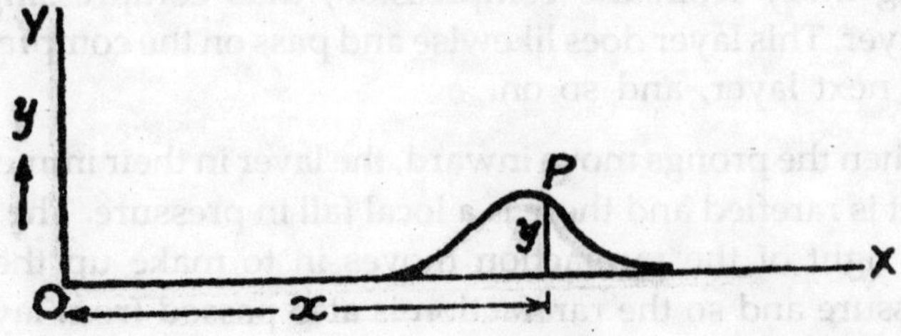

Let the time be measured from the instant when the particle at the origin O is passing through its equilibrium position. Then the displacement y of this particle at any instant t can be represented by

$$y = f(t),$$

where $f(t)$ is any function of time. The wave will reach a point P, distant x from O, in x/v second. Thus the particle at P will start moving x/v second later than the particle at O. Therefore, the displacement y of the particle at P at any instant t will be the same as the displacement at the origin x/v second earlier *i.e.* at $\left(t - \frac{x}{v}\right)$ Hence the displacement at P is given by

$$y = f\left(t - \frac{x}{v}\right)$$

Since v is a constant, we may write it as

$$y = f(vt - x). \qquad \text{...(i)}$$

This is the equation of a wave of *any* shape travelling along the Positive direction of X-axis with a constant velocity v. The function f determines the exact shape of the wave.

Similarly, the equation for a wave travelling along the negative direction of X-axis would be

$$y = f(vt + x). \qquad \text{...(ii)}$$

Thus the functions $f(vt \pm x)$ represent travelling waves. We can also write these functions as $f(x \pm vt)$.

From eq, (i) and (ii), the general equation for the wave can be written as

$$\boxed{y = f_1(vt - x) + f_2(vt + x).} \qquad \text{....(iii)}$$

To get the general differential equation, let us differentiate eq. (iii) twice partially with respect to t. We get

$$\frac{\partial^2 y}{\partial t^2} = v^2 f_1''(v\,t - x) + v^2 f''_2(vt + x) \qquad \text{....(iv)}$$

where f_1 and f_2 are the second differentials of f_1 and f_2 respectively.

Again differentiating eq. (iii) twice partially with respect to x, we get

$$\frac{\partial^2 y}{\partial x^2} = f_1''(v\,t - x) + f_2''(vt + x). \qquad \text{....(v)}$$

Comparing eq. (iv) and (v), we have

$$\boxed{\frac{\partial^2 y}{\partial t^2} = v^2 \frac{\partial^2 y}{\partial x^2}.} \qquad \text{....(vi)}$$

This is the general *differential* equation for a one-dimensional wave.

Equation of a Plane Progressive Harmonic Wave

A plane progressive harmonic wave is the simplest wave in which the particles of the medium perform simple harmonic motion.

Let the source of simple harmonic disturbance be situated at the origin *0*, and the wave travel in the positive direction of the x-axis in a continuous homogeneous medium.

As the wave proceeds, each successive particle of the medium is set in *simple harmonic* oscillation. Following figure shows the displacements *(y)* of the particles against their positions *(x)* at a particular instant.

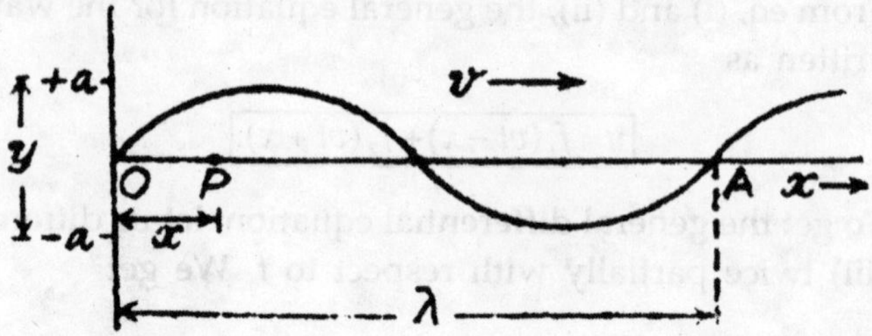

Let the time be measured from the instant when the particle at O is passing through its equilibrium position. Then the displacement of this particle at any instant t is given by

$$y = a \sin \omega t,$$

where a is the amplitude of motion and $\omega = 2\pi$ is the frequency. If the wave velocity is v, then the disturbance produced at O will reach a point P, distant x from O, in x/v second.

Thus the particle at P will start moving x/v second later than the particle at O. Therefore, the displacement of the particle at P at any instant t will be the same as the displacement at O at the instant x/v second earlier, *i.e. at* $\left(t - \frac{x}{v}\right)$. Hence the displacement at P at any instant t can be obtained by substituting $\left(t - \frac{x}{v}\right)$ in place of t in the above equation, and is given by

$$y = a \sin \omega \left(t - \frac{x}{v}\right). \quad \text{...(i)}$$

Now, $\omega = 2\pi / T$, where T is the period of motion. Therefore,

$$y = a \sin 2\pi \left(\frac{\lambda}{T} - \frac{x}{vT}\right)$$

The period T is the time required for the wave to travel a distance of one wavelength λ, so that $vT = \lambda$. Thus

$$y = a \sin 2\pi \left(\frac{\gamma}{T} - \frac{x}{\lambda}\right) \quad \text{...(ii)}$$

This may also be written as

$$y = a \sin \frac{2\pi}{\lambda} \left(\frac{\lambda}{T} t - x\right).$$

But $\lambda / T = v$. Therefore

$$y = a \sin \frac{2\pi}{\lambda} (vt—x). \qquad ...(iii)$$

The quantity $\frac{2\pi}{\lambda} = 2\pi \bar{\nu}$ k is called "propagation constant", where $\bar{\nu} = 1/\lambda$ is the "wave number". Thus

$$y = a \sin k\,(vt-x). \qquad ...(iv)$$

Now $$v = \frac{\lambda}{T} = \lambda \frac{\omega}{2\pi} = \frac{\omega}{k}.$$

$\because$ $$y = a \sin (\omega t - kx). \qquad ...(v)$$

The exponential form of this equation is

$$y = a \exp [\, j(\omega t—kx)]. \qquad ...(vi)$$

Eq. (i).. (ii), (iii), (iv), (v.) and (vi) are the alternative forms of expressing a sinusoidal wave motion of amplitude a, travelling along the $+x$ direction. Here we have assumed that the displacement y is zero at the point x=0 at time t=0. The general expression may be written as, for example,

$$y = a \sin (\omega t—kx—\phi),$$

where ϕ is the phase constant.

These equations show the following properties of wave motion:

(i) At a fixed x, the displacement y is a simple harmonic oscillation in time. It means Miat all particles of the medium oscillate simple harmonically with the same amplitude and period.

(ii) At fixed t, the displacement y is a simple harmonic oscillation in space. It means that at a particular instant the displacement varies sinusoidally from particle to particle, that is, the phase changes from particle to particle.

(iii) If x is increased by λ, then the displacement is given by, from eq. (iii),

$$y' = a \sin\frac{2\pi}{\lambda}\{vt-(x+\lambda)\}$$

$$= a \sin\left\{\frac{2\pi}{\lambda}(vt-x)-2\pi\right\}$$

$$= a \sin\frac{2\pi}{\lambda}(vt-x)\ [\because \sin(\alpha-2\pi)=\sin\alpha]$$

$$= y.$$

Thus, at any time the displacement of the particles at x, $x+\lambda, x+2\lambda$ .....is same. Hence such particles are in the same phase of oscillation.

(iv) Let t increase by δt and x by $v\,\delta t$. Then the displacement is given by, from eq. (iii),

$$y' = a \sin\frac{2\pi}{\lambda}\{v(t+\delta t)-(x+v\,\delta t)\}$$

$$= a \sin\frac{2\pi}{\lambda}(vt-x)$$

$$= y.$$

It means that the disturbance at a point x at an instant t is transferred to another point at a distance $v\,\delta t$ away in time δt. That is, the disturbance is advancing with a velocity v without suffering any change in shape.

(v) The particle velocity is given by, differentiating eq. (iii) with respect to t,

$$\frac{\partial y}{\partial t} = \frac{2\pi va}{\lambda}\cos\frac{2\pi}{\lambda}(vt-x) \qquad \text{....(vii)}$$

This shows that, at any instant, the particle velocity varies from particle to particle simple harmonically. Comparison of eq. (iii) and (vii) shows that where the displacement is maximum, the velocity is minimum and vice-versa.

Particle Velocity in Wave Motion

The individual particles which make up the medium do not travel through the medium with the waves. They simply oscillate about their equilibrium positions. The instantaneous velocity of an oscillating particle of the medium, through which a wave is travelling, is known as 'particle velocity.'

Wave Velocity: The velocity with which the disturbance, or planes of equal phase, travel through the medium is called 'wave (or phase) velocity'.

Relation between Particle Velocity and Wave Velocity: Let us consider a plane simple harmonic wave travelling through a string stretched along the x-axis. The wave equation is

$$y = a \sin \frac{2\pi}{\lambda}(vt - x), \quad \text{...(i)}$$

where y is the instantaneous displacement of a particle at the point x of the string, and a, λ and v are respectively the amplitude, length and velocity of the wave.

The instantaneous particle velocity at the point x is $\partial y/\partial t$ and obtained by differentiating eq. (i). Thus

$$\frac{\partial y}{\partial t} = a\left\{\cos\frac{2\pi}{\lambda}(vt - x)\right\}\frac{2\pi v}{\lambda}. \quad \text{...(ii)}$$

Now, differentiating eq. (i) with respect to x, we have

$$\frac{\partial y}{\partial x} = \left\{\cos\frac{2\pi}{\lambda}(vt - x)\right\}\left(-\frac{2\pi}{\lambda}\right) \quad \text{...(iii)}$$

Dividing eq. (ii) by (iii), we get

$$\frac{\partial y/\partial t}{\partial y/\partial x} = -v$$

or

$$\boxed{\frac{\partial y}{\partial t} = -v\frac{\partial y}{\partial x}}. \quad \text{...(iv)}$$

$\partial y / \partial x$ measures the slope of the string at the point x. Thus the particle velocity at x is equal to the wave velocity multiplied by the slope of the string at x.

Differential Equation of Wave Motion: Differentiating eq. (ii), we get

$$\frac{\partial^2 y}{\partial t^2} = -\text{ a } \left\{\sin\frac{2\pi}{\lambda}(vt - x)\right\}\left(\frac{2\pi v}{\lambda}\right)^2. \quad \text{...(v)}$$

Again, differentiating eq. (iii), we get

$$\frac{\partial^2 y}{\partial x^2} = -\text{ a } \left\{\sin\frac{2\pi}{\lambda}(vt - x)\right\}\left(-\frac{2\pi v}{\lambda}\right)^2. \quad \text{...(vi)}$$

From eq. (v) and (vi), we get

$$\boxed{\frac{\partial^2 y}{\partial t^2} = -v^2\frac{\partial^2 y}{\partial x^2}}. \quad \text{...(vii)}$$

This is the required equation.

Pressure Variation in Longitudinal Waves

Waves in a fluid are always Longitudinal. Now when a longitudinal wave-train (such as sound) travels through a medium, the particles of the medium oscillate to and fro about their equilibrium positions in the direction of propagation of the wave.

The phase of oscillation varies from particle to particle. The distances between the particles so alter that at any instant the particles are alternately crowded together and spread out. Therefore, the pressure varies from particle to particle in the medium.

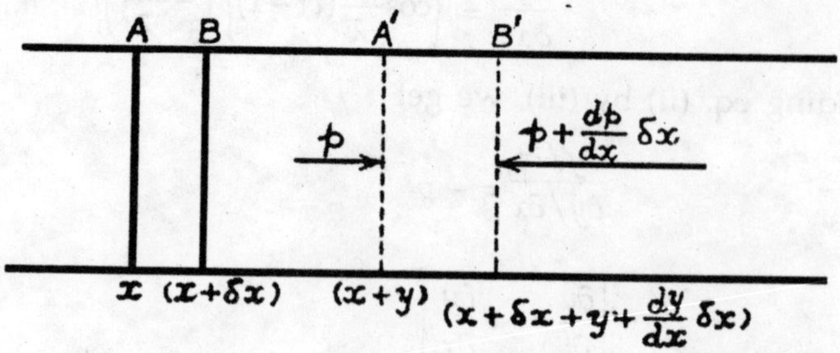

Let us consider a tube of air of unit cross-section with its length along the axis of X. Let *A* and *B* be two planes at right angles to the axis of the tube at distances x and $x+\delta x$ respectively from the origin.

Let a plane longitudinal wave pass along the axis of the tube. Suppose at some particular instant when the wave is passing, the displaced positions of the planes are A' and B' respectively. *Let y* be the displacement of the plane *A*. Then the rate of change of displacement with distance is dy/dx, so that the displacement of the plane *B* is $y+\left(\frac{dy}{dx}\right)\delta x$. Thus the new positions of the planes, A' and B', are $(x+y)$ and $\left(x+\delta x+y+\frac{dy}{dx}\delta x\right)$ respectively.

The area of cross-section of the tube considered is unity. Therefore, the initial volume of air between the planes *A* and *B*

$$= (x+\delta x)-x=\delta x.$$

The final volume of air between the planes (A' and B')

$$= \left(x+\delta x+y+\frac{dy}{dx}\delta x\right) - (x+y)$$

$$= \left(\delta x+\frac{dy}{dx}\delta x\right)$$

$\therefore$ increase in volume

$$= \left(\delta x+\frac{dy}{dx}\delta x\right) - \delta x=\frac{dy}{dx}\delta x.$$

$$\therefore \qquad \text{volume strain} = \frac{\text{increase in volume}}{\text{initial volume}}$$

$$= \frac{\frac{dy}{dx}\delta x}{\delta x} = \frac{dy}{dx}$$

The wave has thus created a volume strain in the air between the planes. Hence, there is a variation of pressure from point to point along the tube.

Now, let p be the pressure-excess above normal on the plane A (now at A'). On B (now at B') the pressure-excess will be p+ $\left(\frac{dp}{dx}\right)\delta x$. These two pressure-excesses will act on the sample of air between A' and B' in the directions shown in figure. These are equivalent to :

(i) Equal and opposite pressure-excesses p, p, which provide the stress on the sample of air between A' and B' and produce the volume strain dy/dx in the sample.

(ii) A resultant pressure-excess $\left(\frac{dp}{dx}\right)\delta x$ in the direction from B' to A' which provides the restoring force causing the sample to come back to its initial position.

Let K be the bulk modulus of air. Then we have, by Hooke's law

$$k = \frac{\text{stress}}{\text{strain}} = -\frac{P}{dy/dx},$$

negative sign showing that when pressure increases, the volume decreases.

$$\therefore \qquad p = -K\frac{dy}{dx}$$

This is the required expression.

dy/dx represents the slope of the displacement curve for the wave. Hence in a longitudinal wave the slope of the displacement curve at any point, dy/dx, measures the pressure-change (compression or rarefaction) at that point. Where the slope dy/dx *is* negative, p is positive (compression) : where dy/dx is positive, p is negative (rarefaction).

Longitudinal Wave as a Pressure Wave

The 'displacement equation' for a plane wave travelling in the $+x$ direction with velocity v is

$$y = a \sin\frac{2\pi}{\lambda}(vt-x), \qquad \text{...(i)}$$

where a is the displacement-amplitude of the wave. In case of longitudinal wave, the displacement y is along the direction of propagation (x-axis) and at any instant the pressure varies from particle to particle in the medium.

Let p be the instantaneous change in pressure at a point x. Then we have

$$P = -K\frac{dy}{dx} \quad \text{...(ii)}$$

where K is the bulk modulus of elasticity of the medium. Now, differentiating eq. (i) with respect to x, we have

$$\frac{dy}{dx} = -\frac{2\pi a}{\lambda}\cos\frac{2\pi}{\lambda}(vt-x).$$

Putting this value of dy/dx in eq. (ii), we get

$$p = \frac{2\pi ka}{\lambda}\cos\frac{2\pi}{\lambda}(vt-x).$$

Now, the longitudinal wave velocity in a medium of bulk elasticity K and density ρ is $v\sqrt{\left(\frac{K}{\rho}\right)}$ =, so that $K = v^2\rho$.

$$\therefore \quad p = \frac{2\pi v^2 \rho a}{\lambda}\cos\frac{2\pi}{\lambda}(vt-x).$$

$2\pi v^2 \rho a/\lambda$ is the maximum value of p (*i.e.* maximum change in pressure) and is called the "pressure-amplitude" p_m. Thus

$$p = p_m \cos\frac{2\pi}{\lambda}(vt-x). \quad \text{...(iii)}$$

Eq. (iii) is the 'pressure equation' of a longitudinal wave. Hence a longitudinal wave, such as a sound wave, may be considered either as a displacement wave [cq. (i)] or as a pressure wave [eq. (iii)]. A comparison of eq. (i) and (iii) shows that the displacement wave is 90° out of phase with the pressure wave. That is, when the displacement at a point is zero, the pressure-change is a maximum, and vice-versa.

The displacement and pressure curves for a longitudinal wave have been drawn in following figure. The pressure curve may be drawn by measuring the slopes, *dy/dx*, of the displacement curve at different points. According to eq. (ii), *dy/dx* at a point *x* measures the pressure-change *p* (compression or rarefaction) at that point. Where the slope *dv/dx* of displacement curve is negative, as from *A* to *B* and from *D* to *E, the* pressure-change is positive which means compression. Where the slope is negative, as from *B* to *D* via *C*, the pressure-change is negative which means rarefaction.

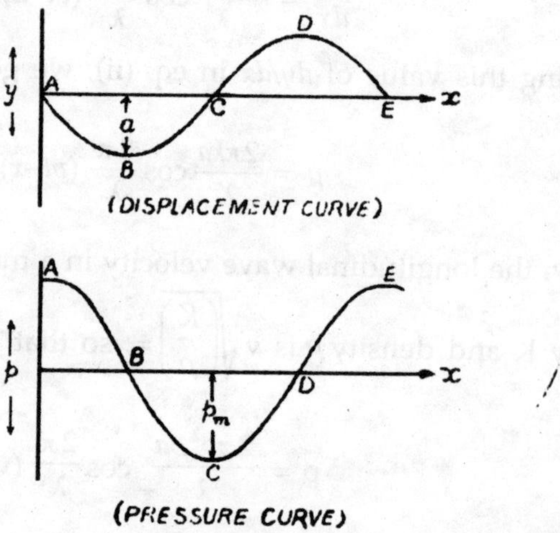

By measuring *dy/dx* at different points *x* on the displacement curve, the corresponding values of $p\left(=-v^2\rho\frac{dy}{dx}\right)$ can be obtained. Then a curve may be plotted between ρ and x. This would be the pressure curve.

Energy Distribution in a Plane Progressive Longitudinal Wave

When a plane harmonic wave, starting from a source of disturbance, travels in a medium, the medium particles are successively set in oscillation. The energy from the source is thus transmitted through the medium in the; direction of the wave. At any instant, the energy of the medium is partly kinetic and partly potential. We have to calculate the total energy per unit volume of the medium.

Suppose the source is situated at the origin and the waves are travelling along the +x direction. An equation of these waves is

$$y = a \sin \omega \left(t - \frac{x}{v}\right). \qquad \text{...(i)}$$

The particle velocity is

$$\frac{dy}{dt} = \omega a \sin \omega \left(t - \frac{x}{v}\right) \qquad \text{...(ii)}$$

Kinetic Energy Density: Let us consider a plane layer of *unit* area and thickness δx of the medium at a distance x from the origin and perpendicular to the direction of wave propagation. The mass of the layer is $\rho\, \delta x$, where ρ is the density of the medium.

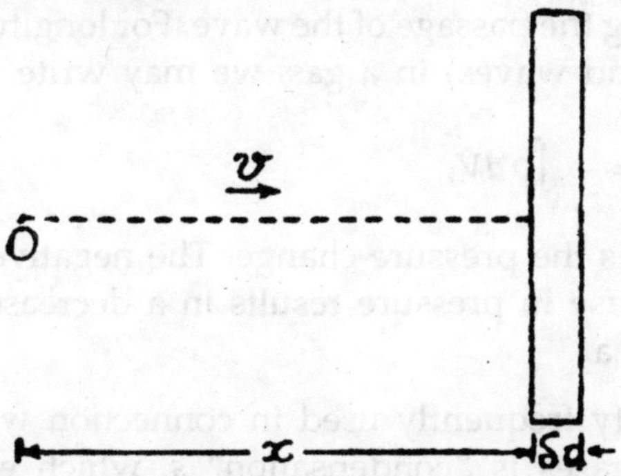

As the layer is thin, we can assume all its particles to have the same velocity. The kinetic energy of the layer is

$$\delta K = \frac{1}{2}(\rho\, \delta x)\left(\frac{dy}{dt}\right)^2 \qquad \text{...(iii)}$$

$$= \frac{1}{2}(\rho\, \delta x)\, \omega^2 a^2 \cos^2 \omega \left(t - \frac{x}{v}\right). \qquad \text{[by eq.(ii)]}$$

The kinetic energy per unit area per unit length, *i.e.* per unit volume can be obtained by taking the average of δK over a wavelength, *i.e.* from $x=0$ to $x=\lambda$. That is

$$\overline{K} = \frac{1}{\lambda}\int_0^\lambda dk$$

$$= \frac{1}{\lambda}\left(\tfrac{1}{2}\rho\omega^2 a^2\right)\int_0^\lambda \cos^2\omega\left(t-\frac{x}{v}\right)dx$$

$$= \frac{1}{\lambda}\left(\tfrac{1}{2}\rho\omega^2 a^2\right)\int_0^\lambda \tfrac{1}{2}\left\{1+\cos 2\omega\left(t-\frac{x}{v}\right)\right\}dx.$$

The value of the integral of the "cos" term is zero. Therefore

$$\overline{K} = \frac{1}{\lambda}\left(\tfrac{1}{2}\rho\omega^2 a^2\right)\int_0^\lambda \tfrac{1}{2}\,dx = \tfrac{1}{4}\rho\,\omega^2 a^2 \qquad \text{...(iv)}$$

Potential Energy Density: The potential energy of a fixed mass of the medium of volume V is equal to the work done upon it during the passage of the wave. For longitudinal waves (such as sound waves) in a gas, we may write

$$\delta U = -\int p\,dV, \qquad \text{...(v)}$$

where p is the pressure-change. The negative sign shows that an increase in pressure results in a decrease in volume, and vice-versa.

A quantity frequently used in connection with vibrating columns of gases is "condensation" s, which expresses the fractional change of density of the gas with respect to the original density. Thus

$$s = \frac{\delta\rho}{\rho} = -\frac{\delta V}{V} = -\frac{dy}{dx},$$

because we know that in longitudinal waves, dv/dx is the volume strain $\delta V/V$.

Let us again consider the layer of unit area and thickness δx. If V be the volume of this layer, then the condensation is

$$S = -\frac{\int dV}{V}$$

where $\int dV$ is the incremental change in the original volume V.

From its differentiation, we have

$$dV = -V\,ds. \qquad \text{...(vi)}$$

Also, for longitudinal waves in a fluid, the pressure-change is given by

$$\rho = -K_v \frac{dy}{dx} = K_v s \qquad \text{...(vii)}$$

where K_v is the volume elasticity of the medium. Substituting for ρ and dV from eq, (vii) and (vi) in eq. (v), we get

$$\delta U = K_v V \int_0^s s\,ds = \frac{1}{2} K_v V s^2$$

Now, $V = \delta x$ (volume of the layer of unit area and thickness δx) and $s = -dy/dx$.

$$\therefore \qquad \delta U = \frac{1}{2} K_v\ \delta x \left(\frac{dy}{dx}\right)^2$$

But we know that for a wave, $\frac{dy}{dx} = -v\frac{dy}{dx}$, where v is the wave velocity. Also for a longitudinal wave, $v = \sqrt{\left(\frac{K_v}{\rho}\right)}$, where ρ is the density of the medium. Thus $K_v = v^2 \rho$. Making these two substitutions in the last expression, we get

$$\delta U = \frac{1}{2}\rho\delta x\left(\frac{dy}{dt}\right)^2,$$

which is the same as the kinetic energy SK of the layer (eq. iii).

Its value over a wavelength gives the potential energy per unit volume. Thus, as in case of kinetic energy,

$$\overline{U} = \tfrac{1}{4}\rho\omega^2 a^2. \qquad \text{...(viii)}$$

Thus, in a plane progressive longitudinal wave the kinetic and the potential energy densities are equal at any point and at any time.

Total Energy Density: The total energy per unit volume, *i.e.* energy density $= \overline{K} + \overline{U}$

$$= \tfrac{1}{4}\rho\omega^2 a^2 + \tfrac{1}{4}\rho\omega^2 a^2 = \tfrac{1}{2}\rho\omega^2 a^2.$$

Since $\omega = 2\pi n$, where n is frequency, this can also be written as

$$\boxed{\text{energy density} = 2\pi^2 n^2 a^2 \rho.}$$

Energy Current or Wave Intensity: All progressive waves transmit energy. The energy flowing per second per unit area perpendicular to the direction of wave propagation is called "energy flux" energy current' or "wave intensity"

Now, the energy density is the energy contained in a cylinder of unit length and unit area of cross-section. The wave travels a distance v per second. Therefore the energy flowing per second per unit area of cross-section will be the energy contained in a cylinder of length v and unit area of cross-section. Thus wave intensity (energy current)= energy density x wave velocity

or $$\boxed{I = 2\pi^2 n^2 a^2 \rho v.}$$

Thus the wave intensity is directly proportional to the density of the medium and the wave velocity, and the square of the amplitude and the square of the frequency.

Pressure Amplitude of a Longitudinal Wave

A plane progressive longitudinal wave moving along the $+x$ direction is represented by an equation

$$y = a \sin \frac{2\pi}{\lambda}(vt - x), \qquad \text{...(i)}$$

where a is the 'displacement' amplitude and v the wave velocity.

During the passage of a longitudinal wave, the pressure varies from point to point in the medium. The pressure change at any point x is given by

$$p = -K\frac{dy}{dx}, \quad \text{...(ii)}$$

where K is the volume elasticity of the medium. Differentiating eq. (i) with respect to x and putting the value of dy/dx so obtained in eq. (ii), we get

$$p = -K\left(\frac{-2\pi a}{\lambda}\right)\cos\frac{2\pi}{\lambda}(vt-x).$$

If ρ be the density of the medium, then $v=\sqrt{K/\rho}$, so that $K=v^2\rho$. Making this substitution in the last expression, we get

$$p = \frac{2\pi v^2 \rho a}{\lambda}\cos\frac{2\pi}{\lambda}(vt-x).$$

The maximum value of the "cos" term is 1. Therefore, the maximum value of ρ, which is the pressure-amplitude, is given by

$$\boxed{p_m = \frac{2\pi v^2 \rho a}{\lambda} = 2\pi n\rho v.} \qquad \left[\because \frac{v}{\lambda} = n\right]$$

This is the required expression.

Intensity and Pressure-amplitude: We know that the intensity of a wave is given by

$$I = 2\pi^2 n^2 a^2 \rho v.$$

This may be written as

$$I = \frac{4\pi^2 n^2 a^2 \rho^2 v^2}{2\rho v}$$

or

$$\boxed{I = \frac{\rho m^2}{2\rho v}}$$

This is the intensity expression in terms of pressure-amplitude p_m.

Problems

1. Which of the following are solutions of the one-dimensional wave equation? (i) $y= 4 \sin x \cos vt$, (ii) $y=9 \sin 2x \cos vt$, (iii) $y=x^2—v^2t^2$, (iv) $y=2x—5t$.

Solution: The general differential equations for a progressive wave $y = f(vt \pm x)$ are :

$$\frac{\partial y}{\partial t} = - v \frac{\partial y}{\partial x} \qquad ...(1)$$

$$\frac{\partial^2 y}{\partial t^2} = - v^2 \frac{\partial^2 y}{\partial x^2} \qquad ...(2)$$

where v is the constant wave-velocity. Let us examine the given equations :

(i) $$y = 4 \sin x \cos vt. \qquad ....(i)$$

Differentiating it twice partially with respect to t and x separately, we get

$$\frac{\partial^2 y}{\partial t^2} = 4 \sin x \; (- v^2 \cos vt) = - v^2 y$$

and $$\frac{\partial^2 y}{\partial t^2} = 4 \sin x \; \cos vt = - y$$

$\therefore$ $$\frac{\partial^2 y}{\partial t^2} = - v^2 \frac{\partial^2 y}{\partial x^2}.$$

This is same as eq. (2) above. Hence the given eq. (i) is a solution of wave equation.

(ii) $$y = 9 \sin 2x \cos vt: \qquad ...(ii)$$

$$\frac{\partial^2 y}{\partial t^2} = 9 \sin 2x \; (- v^2 \cos vt) = - v^2 y$$

and $$\frac{\partial^2 y}{\partial x^2} = - 36 \sin 2x \; \cos vt = - 4y$$

$$\therefore \quad \frac{\partial^2 y}{\partial t^2} = -\frac{v^2}{4}\frac{\partial^2 y}{\partial x^2}.$$

This is different from eq. (2). Therefore, eq. (ii) is *not* a solution of wave equation.

(iii) $$y = x^2 - v^2t^2 \qquad ..(iii)$$

$$\frac{\partial^2 y}{\partial t^2} = -2\,v^2$$

and $$\frac{\partial^2 y}{\partial x^2} = 2$$

$$\therefore \quad \frac{\partial^2 y}{\partial t^2} = -\,v^2\,\frac{\partial^2 y}{\partial x^2}.$$

Clearly, eq. (iii) also is *not* a solution of wave equation.

(iv) $$y = 2x - 5t. \qquad ...(iv)$$

This equation can be differentiated only once.

$$\frac{\partial y}{\partial t} = -5$$

$$\frac{\partial y}{\partial x} = 2.$$

$$\therefore \quad \frac{\partial y}{\partial t} = -\frac{5}{2}\,\frac{\partial y}{\partial t}$$

This is similar to eq. (1), if v=5/2. Hence the given equation (iv) is a solution of wave equation.

2. *What is meant by a propagation vector in a general plane harmonic wave ? If the propagation constant of a wave is 132 per cm and its velocity is 300 metres/sec, which, most be its wave number, wavelength and frequency?*

Solution: The equation of a plane harmonic wave moving along +x–axis, in terms of the propagation constant k is

$$y = a \sin(\omega t - kx).$$

· A physical displacement y may not be associated with all waves. The common feature of all waves is the transmission of some sort of disturbance with a velocity characteristic of the medium. For example, y may correspond to pressure-change in sound waves or electric (or magnetic) field in electromagnetic waves. The disturbance may be represented by a general symbol *ip,* so that we may write

$$\psi = a \sin(\omega t—kx).$$

The equation for a three-dimensional wave would be

$$\psi = a \sin(\omega t - k_x x - k_y y - k_z z)$$

$$\psi = a \sin(\omega t - \vec{k} \cdot \vec{r})$$

where the vector $\vec{k}$ (= $k_x \hat{i} + k_y \hat{j} + k_z \hat{k}$) is called the 'propagation vector.' The direction of this vector is the direction of wave motion and its magnitude is $2\pi/\lambda$, the propagation constant k.

Now, propagation constant $k = \frac{2\pi}{\lambda}$ 132 per cm.

$$\therefore \text{ wave number } \bar{\nu} = \frac{1}{\lambda} = \frac{k}{2\pi} = \frac{132}{2\times(22/7)} = 21 \text{ per cm.}$$

$$\text{Wavelength, } \lambda = \frac{1}{v} = \frac{1}{21} = 0.0476 \text{ cm.}$$

$$\text{Frequency, } n = \frac{v}{\lambda} = v\,\bar{\nu} = 30000\ \frac{cm}{\text{sec}} \times \frac{21}{cm} = 6.3\times10^5/\text{scc.}$$

3. *Write down the equation of simple harmonic progressive waves having following characteristics: (i) amplitude 0.6 cm, period 0.5 sec, wavelength 36 cm, travelling along positive x-axis ; (ii) amplitude 0.003 cm, frequency 660 Hertz, velocity 330 metre /sec, travelling along negative x-axis. Assume the displacement y=0 at x=0 at t=0.*

Solution: (i) The equation of a simple harmonic progressive wave along+x-axis, in terms of amplitude a, period T and wavelength λ is, when y = 0 at x = 0 and t = 0,

$$y = \text{a} \sin 2\pi \left(\frac{t}{T} - \frac{x}{\lambda}\right)$$

Putting a=0.6 cm, T=0.5 sec and A=36 cm, we get

$$y = 0.6 \sin 2\pi \left(2t - \frac{x}{36}\right) \text{cm.}$$

(ii) Again, the equation of a simple harmonic waveing along –x-axis, in terms of frequency n and velocity v, is

$$y = a \sin \omega \left(t + \frac{x}{v}\right)$$

$$= a \sin 2\pi n \left(t + \frac{x}{v}\right)$$

Putting a=0.003 cm=3×l0^{-5} metre, n=660 sec^{-1} and v=330 metre/sec, we get

$$y = 3\times10^{-5} \sin 2\pi \left(660t + \frac{660x}{330}\right)$$

or $$y = 3\times10^{-5} \sin 2\pi \ (660t + 2x) \text{ metre.}$$

4. *Write the wave function for a sinusoidal wave propagating in +x direction with an amplitude of 0.1 metre, wavelength 2*0 metre and frequency 5 Hz, if the point x=0 has zero displacement at time t=0. What is the wave speed?*

Solution: The equation of a sinusoidal progressive wave along +x-axis is

$$y = a \sin \frac{2\pi}{\lambda} (vt—x).$$

Here a=0.1 m, λ =2.0m and

$$v = n\lambda = 5 \text{ sec}^{-1}\times 2.0 \text{ metre} = 10\text{m/s.}$$

$\therefore$ $$y\ (x,t) = 0.1 \sin \pi \ (10t–x) \text{ metre.}$$

5. Write the equation of a simple harmonic progressive wave of amplitude 0.005 m and frequency 50 Hertz, travelling in the negative direction of x-axis with a velocity of 3000 m/sec.

$$\left[\text{Ans.}\, y = 0.005 \sin 2\pi\left(50t + \frac{x}{60}\right) \text{metre}\right]$$

6. The speed of a wave is 250 metres/sec and its frequency is 500 cycles/sec. What is the distance between two points having a phase difference of 60°; and what is the phase difference between two displacements of a point at times 10^{-3} sec apart?

Solution: $v = n\lambda$

$$\lambda = \frac{v}{n} = \frac{250}{500} = 0.5 \text{ metre.}$$

We know that two particles having a phase difference of 2π, *i.e.* 360° are at a distance λ apart. Therefore, the distance between two particles having a phase difference of 60° is

$$\frac{\lambda}{360^\circ} \times 60^\circ = \frac{0.5}{360^\circ} \times 60^\circ = \frac{1}{12} \text{ m.}$$

Again, the period of the wave is

$$T = \frac{1}{n} = \frac{1}{500} = 2\times 10^{-3} \text{ sec.}$$

We know that the phase difference between two displacements of a point at times T apart is 2π (=360°). Therefore the phase difference between two displacements at times 10^{-3} sec apart is

$$\frac{360^\circ}{T} \times 10^{-3} = \frac{360^\circ}{2\times 10^{-3}} \times 10^{-3} = 180^\circ.$$

7. Two particles of a medium disturbed by the wave propagation are at x_1=0 and x_2=1 cm. The displacements of the particles can be given by the equations:

$$y_1 = 2 \sin 3\pi t$$

$$y_2 = 2 \sin (3\pi t - \pi/8).$$

Determine the frequency, wavelength, wave-velocity and the displacement of the particle at t=1 sec and x=4 cm.

Solution: The particle motion is simple harmonic ($y=a \sin \omega t$). This shows that

$$a = 2\text{cm and } \omega = 3\pi .$$

The frequency is

$$n = \frac{\omega}{2\pi} = \frac{3\pi}{2\pi} = 1.5 \text{ Hz.}$$

We see that two particles at a distance 1 cm apart have a phase difference of $\pi/8$. Therefore, the wavelength, *i.e.* the distance between two particles having a phase difference of 2π is

$$\lambda = \frac{8}{\pi} \times 2\pi = 16 \text{ cm.}$$

The wave velocity is

$$v = n\lambda = 1.5 \text{ sec}^{-1} \times 16 \text{ cm} = 24 \text{ cm/sec.}$$

The wave equation

$$y = a \sin \frac{2\pi}{\lambda}(vt-x)$$

becomes

$$y = 2 \sin \frac{2\pi}{16}(24t-x) \text{ cm.}$$

At $t=1$ sec and $x=4$ cm, we have

$$y = 2 \sin 2.5\pi \text{ cm}$$

$$= 2 \sin\left(2\pi + \frac{\pi}{2}\right) \text{cm}$$

$$= 2 \sin \frac{\pi}{2} = \mathbf{2 \text{ cm}} .$$

8. *The equation of a wave is given by*

$$y = 10 \sin 2\pi\left(\frac{t}{0.04} + \frac{x}{200}\right)$$

where y and x are in cm and t in second. Find the amplitude, wave-length, velocity and the frequency of the wave. Find the velocity of ft point x = 400 cm at t = 0.08 sec.

Solution: Let us compare the given equation with the standard form of the wave equation:

$$y = a \sin 2\pi \left(\frac{t}{T} - \frac{x}{\lambda}\right)$$

We find

amplitude, $a = 10$ cm,

wave length, $\lambda = 200$ cm,

frequency, $n = \frac{1}{T} = \frac{1}{0.04} = 25 \text{ sec}^{-1}$,

and velocity $v = n\lambda = 25 \times 200 = 5000$ cm/sec.

The particle velocity at a point x at time t is

$$\frac{\partial y}{\partial t} = 10\left(\frac{2\pi}{0.04}\right)\cos 2\pi\left(\frac{t}{0.04} - \frac{x}{200}\right)$$

Putting x=400 cm and t=0.08 sec, we get

$$\frac{\partial y}{\partial t} = 10\left(\frac{2\pi}{0.04}\right)\cos 0$$

$$= 500\pi = 500 \times 3.14 = 1570 \text{ cm/sec.}$$

9. The equation of a transverse wave travelling along a rope is given by

$$y = 5 \sin \pi (0.02x - 4.0\,t) \text{ cm.}$$

Find the amplitude, frequency, velocity and wavelength. Calculate the maximum transverse velocity of any particle in the rope.

Solution: $y = 5 \sin \pi (0.02\,x - 4.0\,t)$ cm

$$= 5 \sin 0.02\pi\left(x \frac{4.0}{0.02} t\right) \text{ cm}$$

$$= 5 \sin \frac{2\pi}{100}(x-200\ t) \text{ cm.}$$

$$= 5 \sin \frac{2\pi}{100}(200\ t-x-\pi) \text{ cm.}$$

Comparing it with the standard wave equation

$$y = a \sin \frac{2\pi}{\lambda}(vt-x), \qquad \text{...(i)}$$

we get

amplitude, $a = 5$ cm,

wavelength, $\lambda = 100$ cm,

velocity, $v = 200$ cm/sec,

and frequency, $n = \frac{v}{\lambda} = \frac{200}{100} = 2$ Hz.

The particle velocity is obtained by partially differentiating eq. (i). That is,

$$\frac{\partial y}{\partial t} = \frac{2\pi va}{\lambda} \cos \frac{2\pi}{\lambda}(vt-x)$$

The maximum value of the cos term is 1.

$$\therefore \quad \left(\frac{\partial y}{\partial t}\right)_{max} = \frac{2\pi va}{\lambda}$$

$$= \frac{2 \times 3.14 \times 200 \times 5}{100} = 62.8 \text{ cm/sec.}$$

10. The equation of a wave in a stretched string is

$$y = 3 \sin 2\pi\left(\frac{t}{0.04} - \frac{x}{400}\right)$$

in which lengths are expressed in cm and time in second. Calculate the wavelength, amplitude, frequency and velocity of the wave. What are the maximum velocity and acceleration of a particle in the String?

[Ans. 400 cm, 3 cm, 25 sec^{-1}, 10000 cm/sec, 150 π cm/sec, 7500 π" cm/sec^2]

11. A simple harmonic wave travelling in the +x-direction has an amplitude 2 cm, frequency 75 sec^{-1} and velocity of propagation 45 m/sec. Calculate the displacement, particle velocity and acceleration of a medium particle at x=135 cm from the origin at t=3 sec. Given that for the particle at the origin the displacement is zero at t=0.

Solution: Under the given condition the wave may be represented by

$$y = a \sin \frac{2\pi}{\lambda}(vt—x).$$

Here a=2cm, v=45,00 cm/sec, n=75 sec^{-1}, $\lambda = \frac{v}{n} = \frac{4500}{75}$=60 cm, x=135 cm and t=3 sec.

Therefore, the particle displacement

$$y = 2 \sin \frac{2\pi}{60}(4500\times3 - 135) \text{ cm}$$

$$= 2 \sin \left(\frac{891}{2}\pi\right)$$

$$= 2 \sin \left(445 + \frac{\pi}{2}\right)$$

$$= 2 \sin \frac{\pi}{2} = -2\text{cm}.$$

The particle velocity

$$\frac{\partial y}{\partial t} = \frac{2\pi va}{\lambda} \cos \frac{2\pi}{\lambda}(vt-x).$$

We have seen above that

$$\sin \frac{2\pi}{\lambda}(vt-x) = -1.$$

$$\therefore \qquad \cos\frac{2\pi}{\lambda}(vt-x) = 0.$$

Hence
$$\frac{\partial y}{\partial t} = 0.$$

The particle acceleration

$$\frac{d^2y}{dt^2} = -\frac{4\pi^2 v^2}{\lambda^2} a \sin\frac{2\pi}{\lambda}(vt-x)$$

$$= -\frac{4\pi^2 v^2}{\lambda^2} y$$

$$= -\frac{4\pi^2 \times (4500)^2}{(60)^2} \times (-2)$$

$$= 4.4 \times 10^5 \text{cm/sec}^2.$$

12. *Calculate the ratio of the intensities of the following two wares:*

$$y = 6.0 \sin (0.4\ t - 25x) \text{ cm}$$

and
$$y = 2.5 \sin (3.2\ t - 200x) \text{ cm}.$$

Solution: Let us write the given equations in the form

$$y = 6.0 \sin 2\pi \left(\frac{0.4t}{2\pi} - \frac{25}{2\pi}x\right) \text{cm} \qquad \text{...(i)}$$

and
$$y = 2.5 \sin 2\pi \left(\frac{3.2t}{2\pi} - \frac{200}{2\pi}x\right) \text{cm}. \qquad \text{...(ii)}$$

Comparing with the standard form $y = a \sin 2\pi \left(\frac{t}{T} - \frac{x}{\lambda}\right)$.

we get

$$a_1 = 6.0 \text{ cm}, \ T_1 = \frac{2\pi}{0.4}, \lambda_1 = \frac{2\pi}{25}$$

$$a_1 = 2.5 \text{ cm}, T_2 = \frac{2\pi}{3.2}, \lambda_2 = \frac{2\pi}{200}$$

Now, the wave intensity is given by

$$I = 2\pi^2 n^2 a^2 \rho v,$$

where v is the wave-velocity and ρ is the medium-density. Therefore, the ratio of the intensities of two waves in a medium is

$$\frac{I_1}{I_2} = \frac{n_1 a_1^2 v_1}{n_2^2 a_2^2 v_2}$$

$$= \frac{n_1 a_1^2 \lambda_1}{n_2^2 a_2^2 \lambda_2} \qquad [\because v = n\lambda]$$

$$= \left(\frac{T_2}{T_1}\right)^2 \left(\frac{a_1}{a_2}\right)^2 \left(\frac{\lambda_1}{\lambda_2}\right) \qquad \left[\because n = \frac{1}{T}\right]$$

Putting the above values, we get

$$\frac{I_1}{I_2} = \left(\frac{0.4}{3.2}\right)^2 \left(\frac{6.0}{2.5}\right)^2 \left(\frac{200}{25}\right) = \frac{9}{100}.$$

13. Find the intensity of a progressive wire in a fluid of density 1.2 kg m⁻³. The displacement or die wave is given by

$$y = 10^{-5} \cos 2\pi\ (7 \times 10^3\ t - 20x),$$

where y and x are expressed in metre and t in second.

Solution: Let us compare the given cos equation with the standard form

$$y = a \cos 2\pi \left(\frac{t}{T} - \frac{x}{\lambda}\right)$$

We see that

amplitude, $a = 10^{-5}$ metre,

period, $T = \frac{1}{7000}$ sec

wavelength, $\lambda = \frac{1}{20}$ metre.

frequency, $n = \frac{1}{T} = 7000\ sec^{-1}$

and velocity, $v = n = 7000 \times \frac{1}{20} = 350$ m/s.

Now, the wave intensity in a fluid of density ρ is given by

$I = 2\pi^2 n^2 a^2 \rho v,$

$= 2 \times (3.14)^2 \times (7000\ s^{-1})^2 \times (10^{-5}\ m)^2 \times (1.2\ kg\ m^{-3}) \times (350\ ms^{-1})$

$= 40$ joule/(metre^2-sec)

$= 40$ watt/metre^2.

14. Plane harmonic waves of frequency 500 Hertz are produced in air with displacement-amplitude 0.001 cm. Deduce (i) pressure-amplitude, (ii) energy density and (iii) energy flux in the waves. Take density of air as 1.29 gm/litre and speed of sound in air as 340 m/sec.

Solution. The given datas are:

$n = 500$ Hz, a=0.001 cm=10^{-5} metre, $\rho = 1.29$ gm/litre

$= 1.29 \times 10^{-3}\ gm/cm^3 = 1.29\ kg/m^3$ and $v = 340$ m/s.

(i) The pressure-amplitude is given by

$p_m = 2\pi n a \rho v$

$= 2 \times 3.14 \times (500\ s^{-1})\ (10^{-5}\ m)\ (1.29\ kg\ m^{-3})\ (340\ ms^{-1})$

$= 13.8\ N/m^2$.

(ii) The energy density

$= 2\pi^2 n^2 a^2 \rho,$

$= 2\ (3.14)^2\ (500\ s^{-1})^2\ (10^{-5}\ m)^2\ (1.29\ kg\ m^{-3})$

$= 6.36 \times 10^{-4}\ J/m^3$.

(iii) The energy flux is

$$I = \text{energy density} \times \text{wave velocity}$$

$$= 6.36\times10^{-4} \frac{\text{Joule}}{\text{metre}^3} \times 340 \frac{\text{metre}}{\text{sec}}$$

$$= 0.216 \text{ J/m}^{2-s}$$

$$= 0.216 \text{ W/m}^2.$$

15. In a gaseous medium of density 1.25 kg m^{-3}, waves of angular frequency 10^4 rad/sec propagate with velocity $\frac{1}{3}\times10^3$ m/s. If the intensity of the wave be 10^{-12} watt/m^2, find (i) bulk modulus of the medium, (ii) amplitude of waves and (iii) amplitude of the pressure variation.

Solution: (i) The wave velocity in a gaseous medium of density ρ and bulk modulus K is given by

$$v = \sqrt{\left(\frac{K}{\rho}\right)},$$

so that $K = v^2 \rho$

$$= (\tfrac{1}{3}\times10^3 \text{ ms}^{-1})^2 \, (1.25 \text{ kg m}^{-3})$$

$$= 1.4 \times10^5 \text{ N/m}^2.$$

(ii) The intensity of a wave in a gas is given by

$$I = 2\pi^2 n^2 a^2 \rho v,$$

so that
$$a = \sqrt{\left(\frac{I}{2\pi^2 n^2 \rho v}\right)}.$$

Here $I = 10^{-12}$ W/m^2, $\omega = 2\pi n = 10^4$ rad/sec, $\rho = 1.25$ kg m^{-3} and $V = \frac{1}{3}\times10^3\text{ms}^{-1}$.

$$\therefore a = \sqrt{\frac{10^{-12}}{(\frac{1}{2}\times10^8)(1.25)(\frac{1}{3}\times10^3)}} = 7\times10^{-3} \text{ and}$$

(iii) The pressure-amplitude p_m is related to wave intensity by

$$I = \frac{pm^2}{2\rho v}$$

so that $$p_{m''} = \sqrt{(2\rho v I)}\,.$$

Putting the given values, we get

$$p_m = \sqrt{\ }[2(1.25 \text{ kg m}^{-3}) (\tfrac{1}{3} \times 10^3 \text{ ms}^{-1}) (10^{-12} \text{ W m}^{-2})]$$

$$\simeq 3 \times 10^{-5} \text{ N/m}^2.$$

11

Acoustics of Buildings

Characteristic Features

The acoustic properties of a hall or an auditorium have marked effects on speech and music being delivered in the hall or auditorium. We find that in same halls the speech delivered is quite intelligible to a listener. Such halls are said to be acoustically good. On the other hand, in some halls the speech is unintelligible and confusing. Such halls are said to be acoustically bad.

W. C. Sabine, in 1911, studied the problem of satisfactory speech and music in a hall and he laid down the following essential features for good acoustics :

(i) The sound heard must be sufficiently loud in every part of the hall.

(ii) The 'quality' of the sound must remain unaltered.

(iii) The successive syllables spoken must be clear and distinct; each syllable should die away sufficiently quickly to avoid overlapping with the next syllable.

(iv) There should be no undesirable echoes.

(v) There should be neither any concentration of sound nor any zone of silence in any part of the hall.

(vi) There should be no undue noise in the hall.

To fulfil these acoustical demands, the following conditions must be secured in the design of public halls and auditoriums :

Optimum Reverberation

A common defect in halls and auditoriums is the undue persistence of sound of the speaker or the singer. A short sound made in a hall reaches a listener directly as well as after successive reflections from the walls, ceiling and sound of the hall.

The listener, therefore, receives a series of sounds of diminishing intensity (since part of energy is lost at each reflection) and, instead of a single sound, he hears a roll of sound. This persistence of sound is called 'reverberation'.

Similarly, when a prolonged sound is made in the hall, some of the sound waves reach the listener directly, while others reach after successive reflections.

The listener, therefore, receives an increasing sound intensity. After a few seconds the energy emitted per second from the source becomes equal to the energy absorbed by the walls and other materials in the hall and that escaped through the doors, open windows and ventilators per second.

Then the intensity of sound becomes steady. When the source is cut-off, the sound heard does not stop instantaneously because the ear continues to receive the successive reflections until they fall below the minimum audibility. Again, this gradual dying away of sounp is known as 'reverberation'.

The time taken for the sound to fall from average intensity to below the mimimum audibility (or to fall to 10^{-6} of its initial value) is called the "period or time of reverberation".

If the period of reverberation is large, then a rapidly

following sequence of short sounds, such as syllables in speech, will overlap as shown in the following figure.

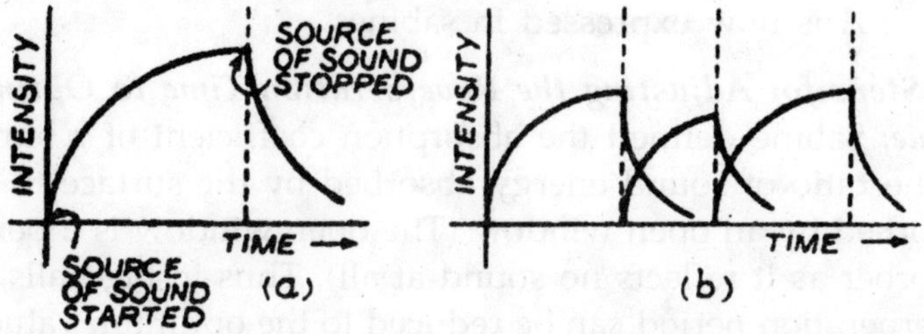

Sabine made an extensive study of the problem of reverberation. He found that there is an optimum period of reverberation depending upon the size of the hall at which best results are obtained. This period is greater for music than for speech.

For example, in a hall of 400,000 ft³ the maximum intelligibility of speech is obtained for a reverberation period between 1·0 and 1.5 seconds, while the music is most appealing between 1.5 and 2·0 seconds. Below this value the intensity of sound is weak and the music appears lifeless, while above this the syllables overlap and become indistinct.

Sabine Formula: Sabine observed that the period of reverberation depends upon the size of the hall and the absorbing power of the objects in it and its walls. He showed that

$$T = \frac{0.162\,V}{A},$$

where T is the period of reverberation in seconds, V is the volume of the hall in metre³ and A is the total absorption of sound in the hall in metric sabines and is given by

$$A = a_1s_1 + a_2s_2 + a_3s_3 + \ldots\ldots$$

where $a_1, a_2, \ldots$ are the absorption coefficients of surfaces of areas $s_1, s_2, \ldots$ (in metre²) respectively of different materials present in the hall.

If V and A are in ft³ and ft², then

$$T = \frac{0.049\,V}{A},$$

where A is now expressed in sabines.

Steps for Adjusting the Reverberation Time to Optimum Value: Sabine defined the absorption coefficient of a surface as the ratio of sound energy absorbed by the surface to that absorbed by an open window (The open window is a perfect absorber as it reflects no sound at all). Thus in big halls, the reverberation period can be reduced to the optimum value by increasing absorption of sound. This is done by taking some of the following steps :

(i) The windows and ventilators are opened.
(ii) Heavy curtains are hung.
(iii) The floor is covered with carpets or tapestries.
(iv) The walls and ceiling are lined with absorbent material such as felt, celotex, fibreboard, glass wool, rock wool, mineral wool, etc.
(v) Acoustic tiles may be provided.
(vi) The clothing worn by the persons in the hall also provides sound absorption. Therefore, the period of reverberation varies according to the number of persons present. To off set this variation, the seats are provided with cushions which themselves have a sufficiently large absorbent effect to make up when the seats are unoccupied.

Adequate Loudness

In a public hall the sounds which reach an audience must be sufficiently loud. This can be achieved by placing large wooden boards near and behind the speaker. Then the sounds reflected from them reach the listener within 1/20 second of the direct sound and increase loudness without creating confusion.

A hard plane wall behind the speaker and facing the audience is also a very good reflecting surface. A low flat ceiling also plays a useful part in reflecting sound towards the audience. If loudspeakers are being used, they should be fitted above the heads of the listeners and pointed slightly downward.

Uniform Distribution

There should be a uniform distribution of the voice of the speaker in the entire hall. For this, the walls and ceiling of the hall should not have curved surfaces because they focus the sound in some parts, while in some other parts no sound reaches at all.

In case curved surfaces are present, they should be covered with absorbent material. A paraboloidal ceiling (the platform being at the focus) is very helpful sending a uniform reflected beam of sound in the hall.

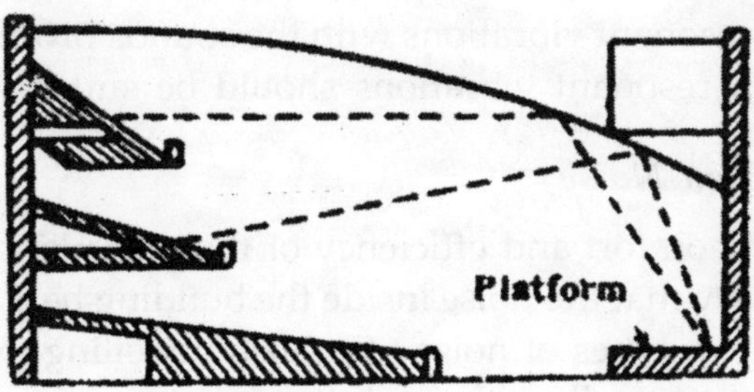

Disturbing Echoes

The disturbing echoes are produced when the time- interval between the sound received by the direct path and that by reflection from some wall or ceilings exceeds about 1/10 sec.

It has been estimated that an interval of 1/20 sec between direct and reflected sounds is the extreme limit of tolerance for speech and 1/15 sec for music. Echoes may be avoided by covering distant walls, and high ceilings with absorbent materials.

Absence of Echelon Effect

An observer walking towards a flight of steps notices that

each footstep is followed by echoes giving a roll of sound. This arises from the regular train of sound reflected from the steps and is called 'echelon effect'. This can be avoided by using a stair carpet.

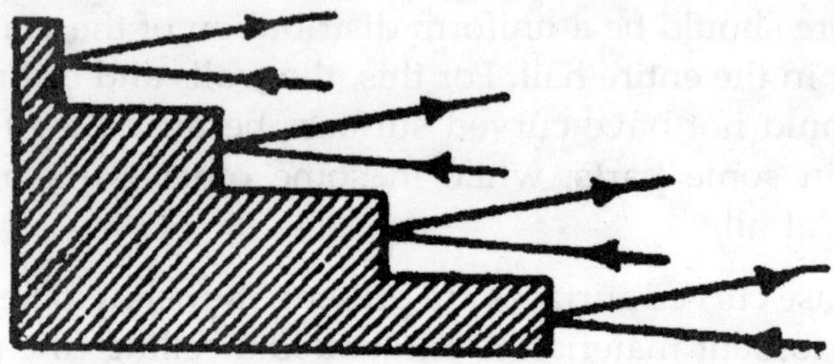

Freedom from Resonance

Sometimes the window-panes, and the wooden panels are thrown in resonant vibrations with the sounds produced in the room. Such resonant vibrations should be suitably damped.

Freedom from Noise

For the comfort and efficiency of those working in offices it is necessary that the noise inside the building be a minimum. There are two types of noise : (i) those travelling through air, and (ii) those travelling through the structure of the building.

The noises travelling through air come in from outside through open windows, doors, etc. This noise can be reduced by using double doors and double or triple windows in separate frames. Noise entering the room through the ventilating ducts can be minimised by packing the ducts with hairfelt baffles or bales of metal gauge. The noise travelling through the walls of the room can be reduced by making the walls of layers of different materials, or by using double walls with air space between them. The noise of foot-steps can be absorbed by carpets or coconut matting.

Noise is also produced inside the room by machinery, type-writers, etc. This may be reduced by proper lubrication of machines, by hanging curtains of absorbent material near the machines, and by placing the typewriters on rubber pads.

Sabine's Reverberation Formula

Sabine obtained from experiments a formula for the time of reverberation of a given room. Franklin in 1903 deduced Sabine's formula theoretically by making the following assumptions :

(i) The distribution of energy is uniform throughout the room.

(ii) There is no interference between sound waves anywhere in the room.

(iii) The absorption coefficient of the surface is independent of the sound intensity.

(iv) The rate of emission of sound energy from the source is constant.

(v) There is no loss of energy in air.

Let us consider an elementary area δS of a plane wall of a room in which sound is being produced. Let us draw two circles with δS as centre, and radii r and $r+\delta r$, in a plane containing the normal δS. The area lying between these circles and enclosed between two radii at angles θ and $\theta+\delta\theta$ with the normal is $r\,\delta\theta \times \delta r$, and is shown shaded.

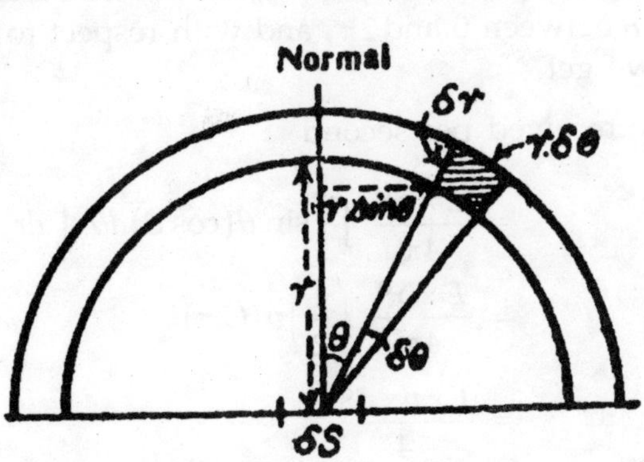

If the whole figure is rotated about the normal through an angle $\delta\phi$, the area considered travels through a circumferential distance r sin $\theta \times \delta\phi$, and traces an element of volume r^2 sin $\theta \times \delta\theta \times \delta r \times \delta\phi$. If be the energy density of sound, then the energy in this volume element at any moment is

$$E \times r^2 \sin\,\theta \times \delta\theta \times \delta r \times \delta\phi\,.$$

This energy is travelling through the element *equally* in all directions. Hence the energy leaving the element through unit solid angle is

$$\frac{E \times r^2 \sin\theta \times \delta\theta \times \delta r \times \delta\phi}{4\pi}$$

The solid angle subtended by δS at the element is $\frac{\delta S \cos\theta}{r^2}$. Therefore, the energy travelling from the element towards δS is

$$\frac{E \times r^2 \sin\theta \times \delta\theta \times \delta r \times \delta\phi}{4\pi} \times \frac{\delta S \cos\theta}{r^2}$$

$$= \frac{E \times \delta S}{4\pi} \times \sin\theta \times \cos\theta \times \delta\theta \times \delta r \times \delta\phi\,.$$

The energy received by δS in one second is obtained by integrating this expression for the whole of the volume lying within a distance of v from δS, where v is the velocity of sound. Integrating with respect to ϕ between 0 and 2π, with respect to θ between 0 and 2π, and with respect to r between 0 and v, we get

energy received per second

$$= \frac{E \times \delta S}{4\pi} \int_0^{\pi/2} \sin\theta(\cos\theta)d\theta \int_0^{v} dr \int_0^{2\pi} d\phi$$

$$= \frac{E \times \delta S}{4\pi}\left(\frac{1}{2}\right) v\ (2\pi)$$

$$= \frac{E \times v \times \delta S}{4}.$$

If a is the absorption coefficient of δS, then

$$\text{energy absorbed per second} = \frac{E \times v \times a \times \delta S}{4}.$$

Therefore, the total energy absorbed per second by the room

$$= \tfrac{1}{4} E \times v \times \sum (a \times \delta S)$$
$$= \tfrac{1}{4} E \times v \times A,$$

where $\sum (a \times \delta S) = A$ is the total absorption of all the surfaces in the room.

Let us now consider the growth and decay of sound energy in the room. Let P be the rate of emission of energy from the sound source, and V the volume of the room. Then the total energy in the room at any instant t is EV, and the rate of its growth is $\frac{d(EV)}{dt} = V\frac{dE}{dt}$. Now,

$$\begin{array}{c} \text{rate of energy} \\ \text{sup ply by sourc} \end{array} = \begin{array}{c} \text{rate of energy} \\ \text{growth in space} \end{array} + \begin{array}{c} \text{rate of energy} \\ \text{absorption by walls} \end{array}$$

$$P = V\frac{dE}{dt} + \frac{1}{4} E\, v\, A$$

or

$$\frac{dE}{dt} + \frac{vA}{4V} E = \frac{P}{V}.$$

Putting $\frac{vA}{4V} = \alpha$, we get

$$\frac{dE}{dt} + \alpha E = \frac{vA}{4V} \alpha.$$

Multiplying by $e^{\alpha t}$, we get

$$\left(\frac{dE}{dt} + \alpha E\right) e^{\alpha t} = \frac{vA}{vV} \alpha\, e^{\alpha t}$$

or

$$\frac{d}{dt}\left(Ee^{\alpha t}\right) = \frac{4P\alpha}{vA} e^{\alpha t}.$$

Integrating, we get

$$Ee^{\alpha t} = \frac{4P}{vA} e^{\alpha t} + C, \qquad ...(i)$$

where C is a constant of integration.

(1) At t=0, E=0. Then from eq. (i), $C=\frac{4P}{vA}$. Therefore, eq. (i) becomes

$$E = \frac{4P}{vA}(1 - e^{-\alpha t}), \text{ where } \alpha = \frac{vP}{4V}$$

or
$$E = E_{max}(1 - e^{-\alpha t}), \text{ where } E_{max} = \frac{4P}{vA}$$

This expression shows the 'growth' of E with time after the start of the sound.

(2) If the source is cut off when E has reached its maximum, then the initial conditions will be

at t=0, P=0 and $E=E_{max}$.

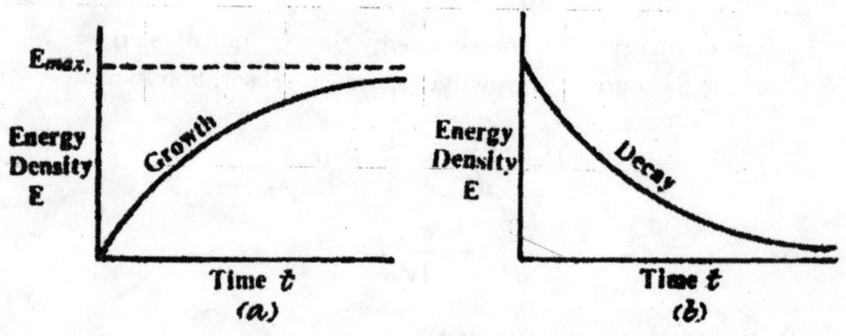

Now, from eq. (i), $C=E_{max}$. Therefore putting P=0 and $C=E_{max}$ in eq. (i), we get

$$Ee^{\alpha t} = E_{max}$$

or
$$E = E_{mvx} e^{-\alpha t}.$$

This expression shows the 'decay' of E with time, after the source has been cut off.

According to Sabine's definition, the reverberation time T is the time during which the steady (maximum) energy density falls to 10^{-6} of its value. Hence putting $t=T$ and $\frac{E}{E_{max}}=10^{-6}$ in the last expression, we get

$$10^{-6} = e^{-\alpha T}$$

or $$\alpha T = \log_e 10^6 = 6 \log_e 10 = 6\times 2.303.$$

$$\therefore T = \frac{6\times 2.303}{\alpha}$$

But $\alpha = \frac{vA}{4P}$

$$\therefore T = \frac{6\times 2.303\times 4V}{vA}.$$

Taking velocity of sound v=340 metre/second at room temperature, we get

$$T = \frac{6\times 2.303\times 4\times V}{340\times A}$$

$$T = \frac{0.162\, V}{A} \text{ second,}$$

where K is the total sound absorption in the room in metric sabine.

If we take sound velocity as 1120 ft/second, then

$$T = \frac{0.049\, V}{A}.$$

This is Sabine's reverberation formula.

This formula does not apply for a 'dead' room in which there is a very large absorption. In the limiting case of complete absorption (a=l), the reverberation timer must be zero. But this result does not follow from the Sabine's formula. The formula is only a 'live' room formula.

Eyring's Formula: Eyring obtained a reverberation formula on the assumption that sound images could be used to replace the walls of a room in calculating the rate of decay of sound energy after the sound source is cut off. His formula reads as

$$T = \frac{KV}{-S \log_e (1-a)}$$

where K is a constant, S is the total area of the absorbing surface, and a is the average absorption coefficient, defined by

$$\bar{a} = \frac{\Sigma(a \times \delta S)}{S} = \frac{A}{S}.$$

This formula applies to 'live' as well as to 'dead' rooms perfectly dead room (open space) $\bar{a}=1$. Then the Eyring's formula gives

$$T = \frac{KV}{-S \log_e 0}$$

$$= \frac{KV}{\infty} \qquad [\because \log_e 0 = -\infty]$$

$$= 0,$$

as expected.

For a 'live' room the absorption coefficient $\bar{a}$ is very small. Then we can write log. (1–a) = – $\bar{a}$. Then the Eyring's formula becomes

$$T = \frac{KV}{S\bar{a}},$$

Echo

The direct reflection of a sound of short duration from a surface of large area such as the wall of a building or a cliff is know as an 'echo'. The best echoes are obtained when the dimensions of the reflector are large compared with the wavelength of the incident sound. Thus a sharp crack on a sound of an impulsive nature gives a loud and clear echo, whereas with a low-pitched sound of long wavelength the echo is hardly perceptible.

The echo is, however, heard if the time-interval between the direct and reflected sound is of the order of 1/10th of a second or more. If the time-interval is less, then a single sharp sound is heard as a roll of sound.

If a sound source be situated at a distance d from a reflecting obstacle, then the time-interval t for the sound to go and return will be given by

$$t = \frac{2d}{v},$$

where v is the velocity of sound. If d= 17m and v=40 m/s, then

$$t = \frac{2 \times 17m}{340 \text{ m/s}} = \frac{1}{10} \text{ second.}$$

If d is less than 17 m, then the time-interval t will be less than 1/10 sec, and the echo cannot be heard.

Problems

1. Define sound absorption coefficient of a material.

A hall of volume 56000 ft³ is found to have a reverberation time of 2 seconds. If the area of sound-absorbing surface be 7000 ft², calculate the average absorption coefficient.

Solution: The sound absorption coefficient of a surface is defined as the ratio of the sound energy absorbed by the surface to that falling on it. Since all the energy falling on an open window passes through it, the absorption coefficient of open window is 1. Hence the absorption coefficient of any surface can also be defined as the ratio of the sound energy absorbed by the surface to that absorbed by an open window.

The reverberation time T is given by

$$T = \frac{0.049\, V}{A}$$

where V the volume of the hall in ft³ and A the total absorption of sound. If S be the total area of the sound absorbing surface and a the average absorption coefficient, then $A = S \times a$. Therefore

$$T = \frac{0.049\, V}{S \times a}$$

Here V= 56,000 ft^3. T=2 sec and S=7000 ft^2.

$$\therefore \quad 2 = \frac{0.049 \times 56{,}000}{7000 \times a}$$

or $$a = \frac{0.049 \times 56{,}000}{2 \times 7000} = 0.196.$$

2. *The reverberation time of a cubical hall of side 30 ft is 0.7 sec. If one of the walls is covered with felt, the reverberation time is reduced to 0.6 sec. Calculate the sound absorption coefficient of felt.*

Solution: The reverberation time T is given by

$$T = \frac{0.049V}{A}. \qquad \text{(Sabine formula)}$$

where V is the volume of the hall in ft^3 and $A = \Sigma as$, a being the absorption coefficient and s the corresponding area.

Here T=0.7 sec, V=(30)^3 ft^3 and A (for the four walls, ceiling and floor)=$a \times (30)^2 \times 6$.

$$\therefore \quad 0.7 = \frac{0.049 \times (30)^3}{a \times (30)^2 \times 6}$$

On solving, we get

$$a = 0.35.$$

This is the absorption coefficient for the walls, ceiling and floor. If one of the walls is covered with felt whose absorption coefficient is a' (say), then

$$A = a \times (30)^2 \times 5 + a' \times (30)^2$$

$$= 0.35 \times (30)^2 \times 5 + a' \times (30)^2.$$

Now the reverberation time is reduced to 0.6 sec. Thus, again from Sabine formula, we have

$$0.6 = \frac{0.049 \times (30)^3}{0.35 \times (30)^2 \times 5 + a' \times (30)^2}$$

$$= \frac{0.049 \times 30}{0.35 \times 5 + a'} = \frac{1.47}{1.75 + a'}$$

On solving, we get

$$a' = 0.70.$$

3. Find the reverberation time of an office which has a volume of 2000 m³ and a total sound absorption of 90 metric Sabine. What is the additional sound absorption required for an optimum reverberation time of 1.5 sec ?

Solution: By Sabine formula, the reverberation time of a room is given by

$$T = \frac{0.162V}{A},$$

where V is the volume of the room in m^3 and A is the total absorption in metric Sabine. Substituting the given values, we get

$$T = \frac{0.162 \times 2000}{90}$$

$$= 3.6 \text{ sec.}$$

The total absorption A for an optimum reverberation T=1.5 sec will be given by

$$A = \frac{0.162\ V}{T}$$

$$= \frac{0.162 \times 2000}{1.5} = 216 \text{ metric sabine.}$$

$\therefore$ additional sound absorption required

$$= 216 - 90 = 126 \text{ metric sabine.}$$

12

Recording and Reproduction of Sound

Recording of Sound on Films

In motion pictures the sound is recorded in the form of a "sound-track" along one edge of the cinema film on which the pictures have been photographed. The sound-track consists of a number of images of a slit. There are two methods of recording:

Variable Area Method: In this method, the width of the sound-track goes on varying with the pressure-variation in the sound wave, the density of the track remaining uniform throughout.

The basic arrangement is shown in the figure given below. It consists essentially of a form of electrical oscillograph. The moving part of the oscillograph is a thin loop of phosphor-bronze suspended in a strong magnetic field parallel to the plane of the loop. The ends of the loop are connected to a microphone through an amplifier.

A small and light mirror is fixed on the loop. A beam of

light, taken from a source of constant intensity and made convergent by a lens, is reflected from the mirror on to a narrow fixed slit. A film is running behind the slit.

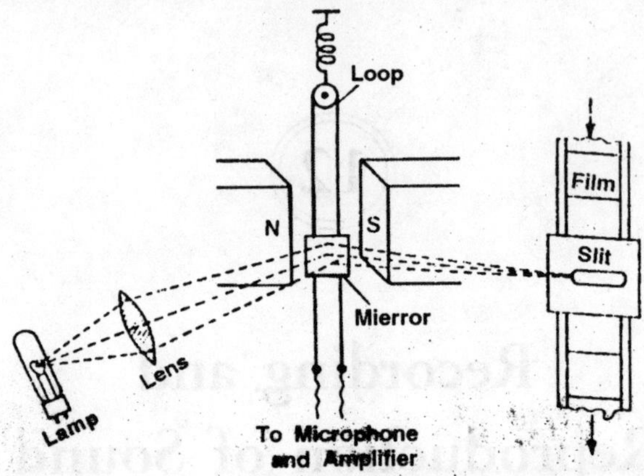

The sound which is to be recorded on the film is sent into the microphone which converts the sound into a fluctuating current. This current is amplified and flows through the loop. Its directions, at any instant, are opposite in the two parts of the loop. So two equal, parallel and opposite mechanical forces at right angles to the plane containing the current and the magnetic field act on the two parts of the loop.

They form a couple and so the loop rotates. Since the current is fluctuating, the plane of the loop, and hence the mirror, has an oscillatory motion. The slit is, therefore, illuminated over varying width. Hence a track of uniform density but variable width is obtained on the film after the film is developed.

(ii) ***Variable Density Method:*** In this method, the whole width of the slit is illuminated but the intensity of light illuminating it varies. Hence the width of the sound track on the film remains uniform throughout but its density varies. This can be done by two methods :

(a) In one method, a special type of discharge lamp

illuminates the slit. The lamp contains an anode plate, a cathode filament coated with barium oxide, and a little helium. The filament is heated by a low-tension battery (L.T.). The plate is connected through a high-tension battery (H.T.) to the secondary *S* of a transformer.

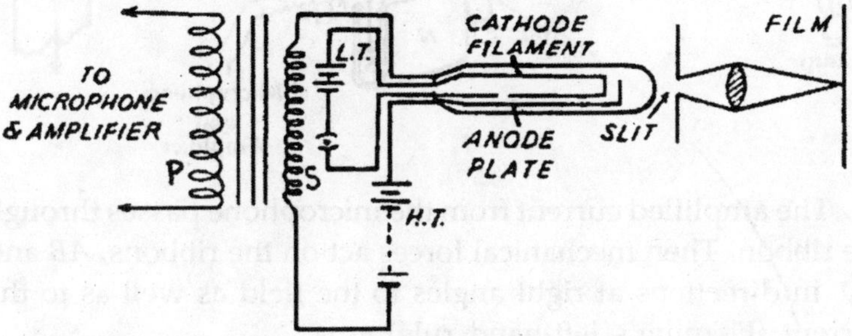

The primary *P* of the transformer is connected to a microphone and amplifier. When the filament is heated, a glow emanates from the anode plate and illuminates the slit. A reduced image of the slit is projected on the moving film by means of a lens.

The sound to be recorded is delivered into the microphone. The microphone current, after amplification, passes through the primary *P* and induces a varying current in the secondary *S*. The secondary current is superposed over the steady plate current. This causes the intensity of the glow, which is illuminating the slit, to vary in accordance to the variations in the microphone current. Hence a track of variable density is photographed on the film.

The disadvantage of this method is that the intensity of light given by the lamp is low and is not strictly proportional to the current through the lamp.

(b) In this method, light from a lamp of constant intensity is first concentrated on a narrow slit cut in an opaque vertical screen, and then focused on the moving film by means of another lens. The slit is kept within a gap between two metal ribbons *AB* and *CD* which lie in the strong magnetic field of a permanent magent at right angles to its plane. The ends of the ribbon are connected to a microphone and an amplifier.

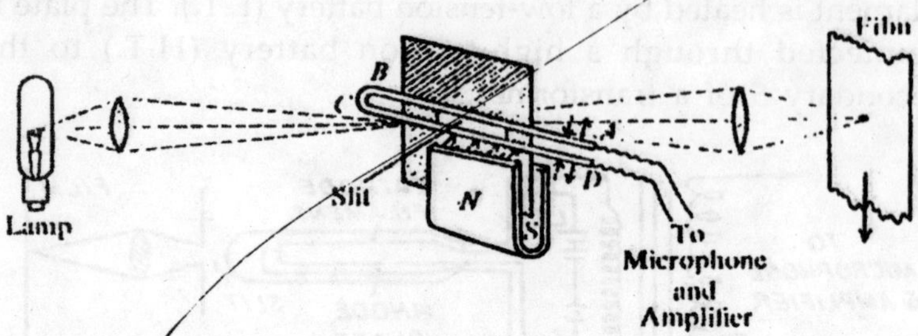

The amplified current from the microphone passes through the ribbon. Then mechanical forces act on the ribbons, *AB* and *CD*, in directions at right angles to the field as well as to the current (Fleming's left-hand rule).

The field is horizontal and the current at any instant is in opposite directions in the two ribbons. Hence the forces will be vertical but oppositely directed as shown by arrows. When the current is reversed, the forces are also reversed.

Thus the width of the gap, and hence of the slit, varies. Consequently the amount of light passing through the slit varies in a similar manner. Thus light of variable intensity falls upon the film on which a track of variable density is obtained.

Reproduction of Sound from Films: The necessary arrangement is shown in the following figure. Light from a powerful source of constant intensity is focused by a condensing lens on a slit whose image is projected on a 'sound gate.' The film moves behind the sound gate at a uniform rate.

The light after passing through the soundtrack on the film falls upon a photoelectric cell. The intensity of the light passing through the sound-track follows the variations in the sound track. Hence the photocell produces minute electric currents which follow the same variations. These are amplified and passed on to a loudspeaker which reproduces the original sound.

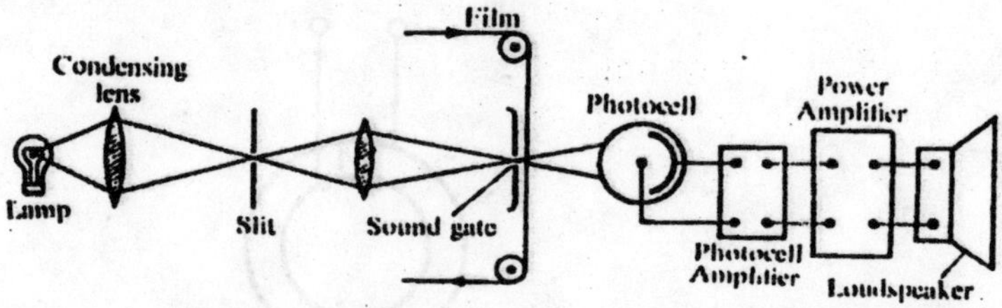

Magnetic Sound Recording

It is the most modern device of sound recording and reproduction, depending upon the magnetic properties of a steel wire or suitably prepared tape.

Principle: Sound waves are collected by a microphone, turned into electric oscillations, amplified and fed to a small electromagnet. The strength of the magnet then varies with the sound waves falling on the microphone. A wire is now moved across the pole of the magnet. The magnetism induced along the length of the wire varies in like manner. Sound waves are thus recorded as magnetic variations on the wire. By a converse process, the magnetic variations are converted into electrical oscillations, which are fed on a loudspeaker and the original sound is reproduced.

Important Parts: There are four main parts of a tape recorder:

(i) *Magnetic Carrier:* It is a magnetisable material in the form of either a stainless steel wire of 0.004" diameter, or a tape consisting of a base of cellulose acetate $\frac{1}{2000}$" thick and $\frac{1}{4}$" wide coated on one side with finely divided ferric oxide. The tape has a high retentivity and sufficient coercivity, i.e. it retains the magnetisation imposed upon it and resists the demagnetising effects.

(ii) *Erasing Head:* It is an electromagnet fed with a powerful alternating current of high frequency (100 kilocycles/sec) from an oscillator. It is used to wipe out any previous recordings on the tape so as to make it usable for a new recording.

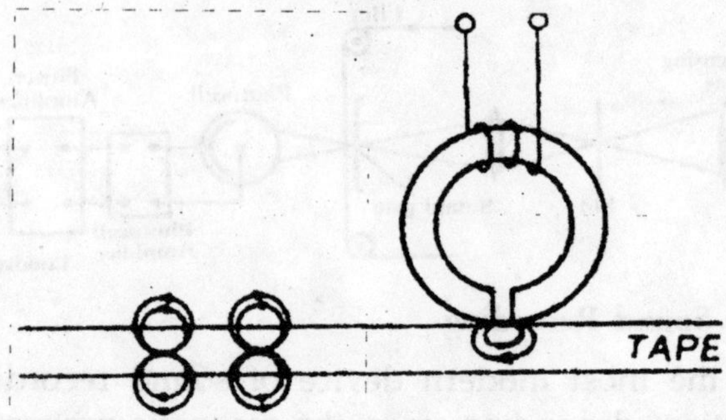

(iii) *Recording Head:* It is another electromagnet having a narrow gap between its pole-pieces and fed by a high-frequency alternating current from the oscillator, plus the amplified current from the microphone. The pole-pieces are tapered to a sharp point to provide a concentrated magnetic field in the gap. It is this head which records the sound message on the tape.

(iv) *Playback Head:* It is also an electromagnet with a core of an extremely high permeability material such as mumetal and permalloy. It converts the varying magnetisation along the tape into electrical oscillations.

Working: The spool on which the tape is wound is rotated by a small electric motor at a constant speed of about 90 metres per minute. The tape then passes across and in intimate contact of the pole-pieces of the erasing head, recording head, playback head, and finally wound on another spool. The whole process consists of three parts :

(i) *Erasing:* As the tape moves across the erasing head which carries a high-frequency alternating current, it passes through a series of hysteresis curves of rapidly diminishing size and is left completely demagnetised.

(ii) *Recording:* The tape next passes across the recording head energised by a high-frequency current, plus the amplified phonic current. The high-frequency current sets the particles of the magnetisable material vibrating and thus makes them more easily influenced by the varying magnetic field produced by the phonic current. Hence the tape acquires residual magnetisation along its length which follows the variations of the phonic current.

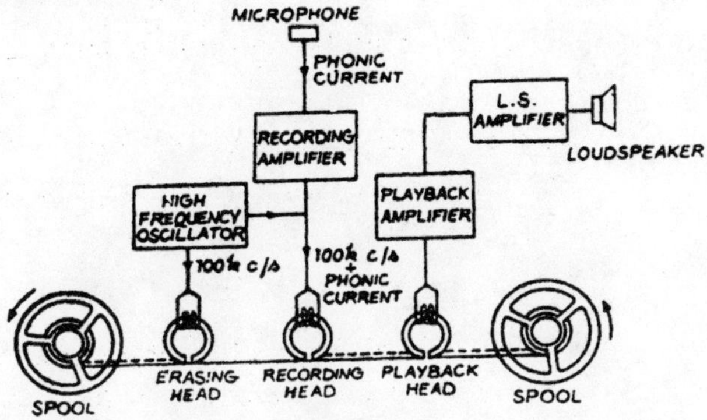

(iii) *Playback:* Finally, the variably-magnetised tape moves across the playback head with the same speed as that across the recording head. The varying magnetisation along the tape causes variations in the magnetic flux through the coil of the playback head. Hence similarly-varying current is induced in the coil which is amplified and fed upon a loudspeaker. The original sound is thus reproduced.

Advantages:

(i) The properties of the tape and the design of the recorder are such as the recording can be reproduced thousands of times without suffering any loss of quality.

(ii) The same tape may be used again and again for fresh recordings.

(iii) The process is quick and handy. Long programmes can be recorded without interruption and reproduced immediately after recording.

(iv) Tape recordings can be preserved for very long periods. It is little affected by mechanical vibration and shock.

(v) The method is widely used in broadcasting, social functions, business, industry and offices.

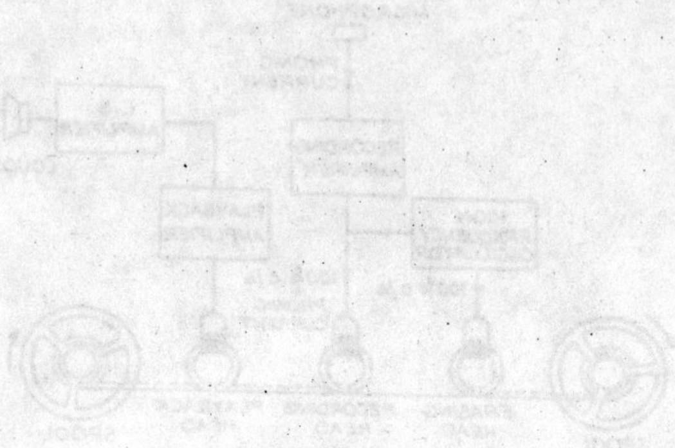

13

Elementary Theory of Filters

Electrical Wave Filters

Electrical wave filters (or simply filters) are electrical networks which transmit electric signals within specified frequency-ranges and suppress signals outside these ranges. An ideal filter would transmit the wanted signals with zero attenuation and suppress the unwanted signals completely.

No actual filter possesses these requirements due to ohmic resistance and terminal impedance inequalities. An ideal filter is, therefore, a network of pure reactances.

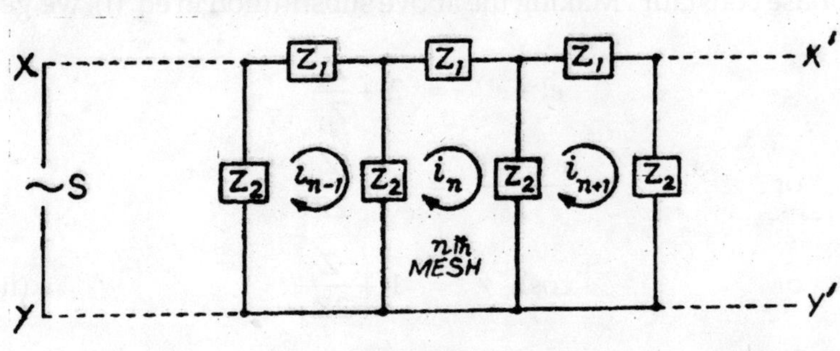

Let us consider a ladder type network consisting of a

number of equal impedances Z_1 in series arm XX' and other equal impedances Z_2 in shunt arms. Let the left end of the network be connected with a varying-frequency voltage source S. Current will flow in each mesh of the network. Let i_{n-1}, i_n and i_{n+1} be the currents in the (n –l)th , nth and (n +l)th mesh respectively.

Applying Kirchhoff's law to the nth mesh, we get

$$i_n Z_1+(i_n-i_{n+1}) Z_2 + (i_n-i_{n-1}) Z_2=0$$

or
$$i_n(Z_1+2Z_2) = Z_2 (i_{n+1}+i_{n-1})$$

or
$$\frac{Z_1+2Z_2}{Z_2} = \frac{i_{n-1}+i_{n+1}}{i_n}$$

or
$$\frac{i_{n-1}}{i_n} + \frac{i_{n+1}}{i_n} = 2+\frac{Z_1}{Z_2}. \qquad \text{...(i)}$$

Due to the impedances there will be a fall in current down the network. Since all the meshes have exactly equal impedances, the current ratio in successive meshes will be same. That is,

$$\frac{i_1}{i_2} = \frac{i_2}{i_3} = ... = \frac{i_{n-1}}{i_n} = \frac{i_n}{i_{n+1}} = ... = e^{\gamma} \text{ (say)},$$

where γ is called the 'propagation constant'. If $\gamma = \alpha + j\beta$, then α is called the 'attenuation constant' and β is called the 'phase constant'. Making the above substitution in eq. (i), we get

$$e^{\gamma} + e^{-\gamma} = 2+\frac{Z_1}{Z_2}$$

or
$$\frac{e^{\gamma} + e^{-\gamma}}{2} = 1+\frac{Z_1}{2Z_2}$$

or
$$\cosh \gamma = 1+\frac{Z_1}{2Z_2}. \qquad \text{...(ii)}$$

If γ is purely imaginary (α =0), that is $\gamma = j\beta$, then there is no attenuation and currents in successive meshes differ in phase only. The eq. (ii) now becomes

$$\cos\beta = 1 + \frac{Z_1}{2Z_2}. \qquad [\because \quad \cosh j\beta = \cos\beta]$$

The limits of $\cos\beta$ are ±1. Therefore, the value of $1 + \frac{Z_1}{2Z_2}$ lies between +1 and –1, that is,

$\frac{Z_1}{Z_2}$ lies between 0 and –4.

Thus the currents (electrical signals) of frequencies for which the impedance ratio Z_1/Z_2 lies between 0 and –4 will pass through the network without attenuation, while the others will be lost in transit. By suitably choosing the values of Z_1 and Z_2, it is possible to select particular frequency-range for transmission. The network then acts as a filter.

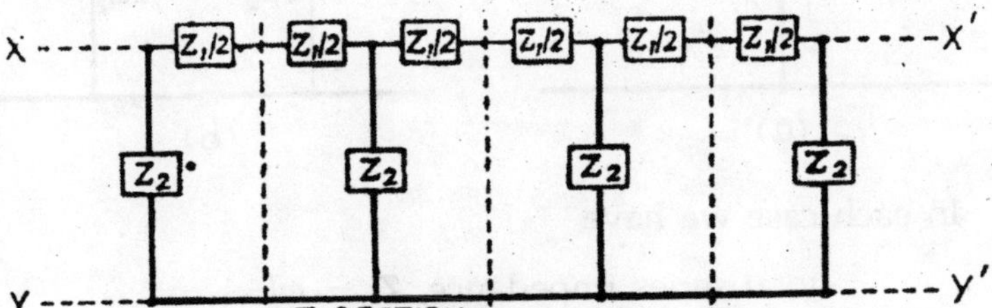

(a)

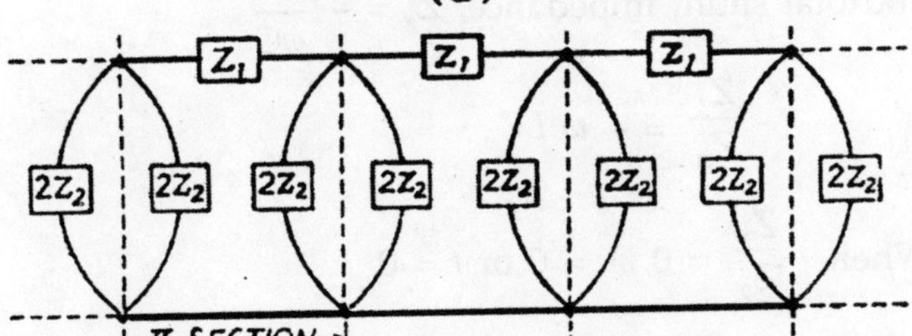

(b)

Types of Filters: Electric wave-filters are classified according to their frequency characteristics which depend on the choice of 2 and Z_2. There are four important types :

1. Low-pass filters.
2. High-pass filters.
3. Band-pass filters.
4. Band-stop filters.

Low-pass Filter: A low-pass filter is one which passes all frequencies between zero (d.c.) and a certain limiting frequency, called the cut-off frequency f_c, and stops all those above f_c. For such a filter, the series arm of the network consists of inductances only and the shunt arms consist of capacitance only.

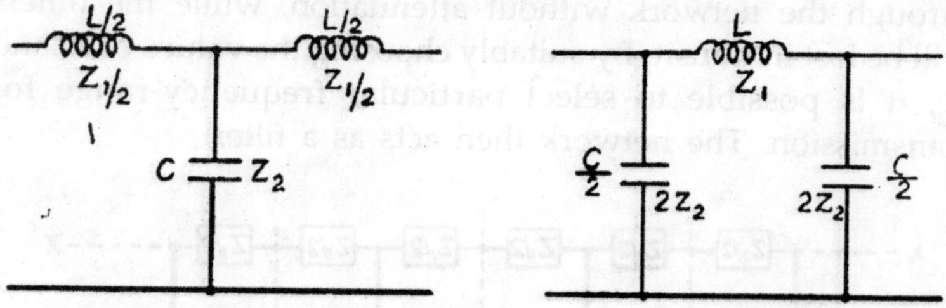

(a) (b)

In each case we have

$$\text{total series impedance, } Z_1 = \omega L$$

$$\text{and total shunt impedance, } Z_2 = -\frac{1}{\omega C}.$$

$$\therefore \quad \frac{Z_1}{Z_2} = -\omega^2 LC.$$

When $\dfrac{Z_1}{Z_2} = 0, \omega = 0$ or $f = 0$

and when $\dfrac{Z_1}{Z_2} = -4, \omega = \dfrac{2}{\sqrt{LC}}$ or $f_c = \dfrac{\omega}{2\pi} = \dfrac{1}{\pi\sqrt{LC}}$.

Thus the filter passes all frequencies between zero and $f_c = \dfrac{1}{\pi\sqrt{LC}}$ and attenuates all others beyond this range.

High-pass Filter: A high-pass filter is one which passes all frequencies about a certain cut-off frequency f_c upto infinity, and stops all those below f_c. For such a filter, the series arm of the network consists of capacitance only and the shunt arms consist of inductances only.

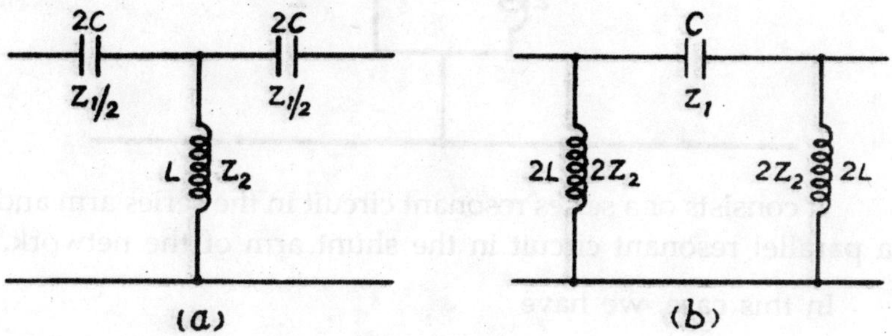

In each case, we have

$$Z_1 = -\frac{1}{\omega C}$$

and $$Z_2 = \omega L$$

$$\therefore \quad \frac{Z_1}{Z_2} = -\frac{1}{\omega^2 LC}.$$

When $\dfrac{Z_1}{Z_2} = 0, \omega = \infty$ or $f = \infty$

and when $\dfrac{Z_1}{Z_2} = -4, \omega = \dfrac{2}{2\sqrt{LC}}$ or $f_c = \dfrac{\omega}{2\pi} = \dfrac{1}{4\pi\sqrt{LC}}.$

Thus the filter passes all frequencies from $f_c = \dfrac{1}{4\pi\sqrt{LC}}$ to infinity, and cuts-off all lower frequencies.

Band-pass Filter: A band-pass filter is one which passes a given range of frequencies, and attenuates all other frequencies on both sides of this pass band.

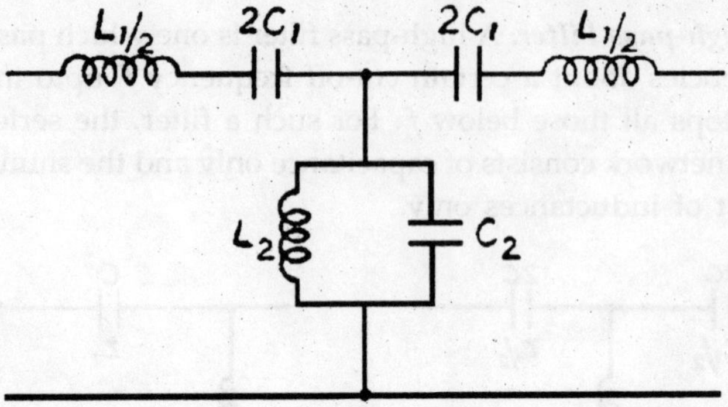

It consists of a series resonant circuit in the series arm and a parallel resonant circuit in the shunt arm of the network.

In this case, we have

$$Z_1 = \omega L_1 - \frac{1}{\omega C_1}$$

and $$\frac{1}{Z_2} = \frac{1}{\omega L_2} - \omega C_2.$$

$$\therefore \quad \frac{Z_1}{Z_2} = \left(\omega L_1 - \frac{1}{\omega C_1}\right)\left(\frac{1}{\omega L_2} - \omega C_2\right).$$

When $$\frac{Z_1}{Z_2} = 0, \omega = \frac{1}{\sqrt{(L_1 C_1)}} = \frac{1}{\sqrt{(L_2 C_2)}}$$

or $$f = \frac{1}{2\pi\sqrt{(L_1 C_1)}} = \frac{1}{2\pi\sqrt{(L_2 C_2)}}. \quad \text{...(i)}$$

Thus $$L_1 C_1 = L_2 C_2.$$

This is in fact the resonant frequency of each circuit.

The cut-off frequencies are given by

$$\frac{1}{Z_2} = -4$$

or $$\left(\omega L_1 - \frac{1}{\omega C_1}\right)\left(\frac{1}{\omega L_2} - \omega C_2\right) = -4$$

or $$\frac{\omega^2 L_1 C_1 - 1}{\omega C_1}\ \frac{1-\omega^2 L_2 C_2}{\omega L_2} = -4$$

or $$(\omega^2 L_1 C_1 - 1)\ (\omega^2 L_2 C_2 - 1) = 4\,\omega^2 L_2 C_1$$

or $$(\omega^2 L_1 C_1 - 1)^2 = 4\,\omega^2 L_2 C_1 \quad [\because L_1 C_1 = L_2 C_2]$$

or $$\omega^2 L_1 C_1 - 1 = \pm\ 2\omega\sqrt{(L_2 C_1)}$$

or $$\left.\begin{aligned} &\omega^2 L_1 C_1 + 2\omega\sqrt{(L_2 C_1)} - 1 = 0 \\ \text{and}\quad &\omega^2 L_1 C_1 - 2\omega\sqrt{(L_2 C_1)} - 1 = 0. \end{aligned}\right\}$$

Solving each of these equations and taking only the positive values of ω, we get

$$\omega_1 = \frac{-\sqrt{(L_2 C_1)} + \sqrt{(L_2 C_1 + L_1 C_1)}}{L_1 C_1}$$

and $$\omega_2 = \frac{-\sqrt{(L_2 C_1)} + \sqrt{(L_2 C_1 + L_1 C_1)}}{L_1 C_1}$$

Thus the two cut-off frequencies are

$$f_{c1} = \frac{\sqrt{[(L_1 + L_2) C_1]} - \sqrt{(L_2 C_1)}}{2\pi L_1 C_1}. \qquad \text{...(ii)}$$

$$f_{c2} = \frac{\sqrt{[(L_1 + L_2) C_1]} - \sqrt{(L_2 C_1)}}{2\pi L_1 C_1}. \qquad \text{...(iii)}$$

Thus the band-pass filter passes all frequencies lying between f_{c1}, ana f_{c2}, given above, and attenuates all those lying outside this range

We can easily show from eq. (i), (ii) and (iii) that

$$f = \sqrt{(f_{c1}f_{c2})}.$$

That is, the resonant frequency of each arm of the network is the geometric mean of the two cut-off frequencies.

Band-stop Filter: A band-stop filter is one which stops a given band (range) of frequencies and passes all the rest. It consists of a parallel resonant circuit in the series arm and a series resonant circuit in the shunt arm of the network, as shown in

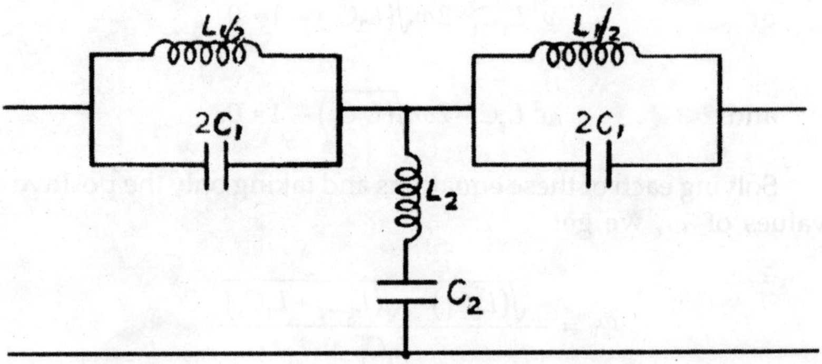

We can show that this filter passes all those frequencies which lie *outside* the two cut-off frequencies, given by

$$f_{c1} = \frac{\sqrt{[L_1(16C_1 + C_2)]} - \sqrt{(L_1C_2)}}{2\pi(4L_1C_1)}$$

and

$$f_{c2} = \frac{\sqrt{[L_1(16C_1 + C_2)]} + \sqrt{(L_1C_2)}}{2\pi(4L_1C_1)}.$$

Acoustic Filters

An acoustic filter is a device which filters out (attenuates) certain frequencies from a complex wave passing through it and allows the others to get through with negligible attenuation.

Its action is based on the acoustic properties of the tubes of different dimensions having cavities and necks.

There are many types of filters according to their frequency characteristics. Let us consider three important types :

Low-pass Filters: In the electric low-pass filters the series elements consist of 'inductances' only while the shunt elements consist of 'capacitance' only.

An acoustic low-pass filter consists of a tube having branch cavities of considerable size arranged close together. The parts of the tube serve as *L* and cavities as C.

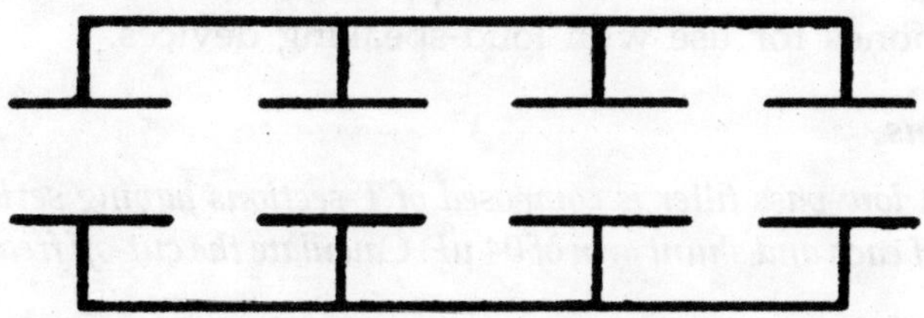

High-pass Filters: In electric high-pass filters the series elements consist of 'capacitance' only while the shunt elements consist of inductances' only.

An acoustic high-pass filter is made of a straight wide tube having holes which carry short necks (L) surrounding them.

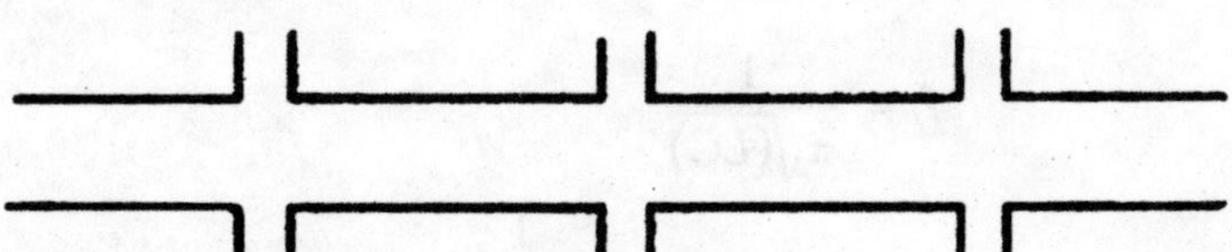

Band-pass Filters: An acoustic band-pass filter is a combination of the other two types. Its main tube has side holes with short necks which lead to closed vessels of considerable volume.

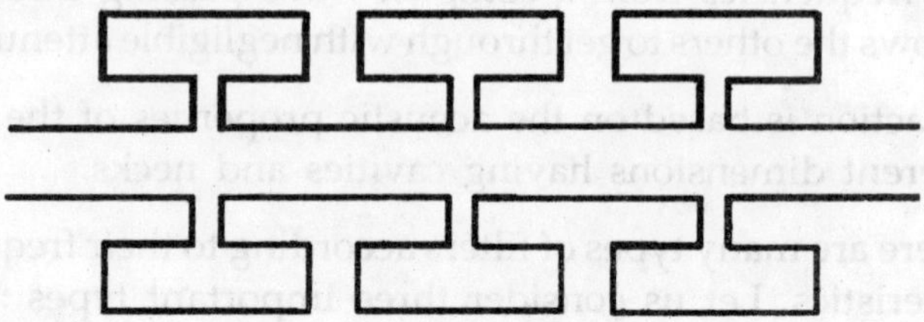

Applications of Sound Filters: Sound filters are used to eliminate undesirable components from a sound wave. For instance, a low-pass filter in the "tone-arm" of a gramophone may filter out the noise due to "scratch" of the needle.

Further, the filters make possible the simultaneous reception of an indefinite number of radio messages with the same antenna. They have also an application in the design of megaphones for use with loud-speaking devices.

Problems

1. A low-pass filter is composed of T-sections having series arm of 20 mH each and shunt arm of 04 μF. Calculate the cut-off frequency.

Solution:

$$\frac{L}{2} = 20 \text{ mH, so that } L = 40 \text{ mH} = 40\times10^{-3} \text{ H}$$

and $$C = 0.4 \ \mu\text{F} = 0.4\times10^{-6} \text{ F.}$$

The cut-off frequency of a low-pass filter is given by

$$f_c = \frac{1}{\pi\sqrt{(LC)}}$$

$$= \frac{1}{3.14\sqrt{\left[(40\times10^{-3})(0.4\times10^{-6})\right]}}$$

$$= 2.5\times10^3 \ c/s = 2.5 \text{ kc/s.}$$

2. *A low-pass filter is composed of T-sections having series arm of 63.5 mH inductance and shunt arms of 0.088 pF each. Compute the cut-off frequency.*

Solution:

$$L = 63.5 \text{ mH} = 63.5 \times 10^{-3} \text{ H}$$

and $$\frac{C}{2} = 0.088 \ \mu\text{F, so that } C = 0.176 \times 10^{-6} \text{ F.}$$

The cut-off frequency is

$$f_c = \frac{1}{\pi\sqrt{(LC)}}$$

$$= \frac{1}{3.14\sqrt{\left[\left(63.5\times10^{-3}\right)\left(0.176\times10^{-6}\right)\right]}}$$

$$= 3.0 \times 10^{3} \ c/s = 3.0 \text{ kc/s.}$$

3. *A high-pass filter is composed of T-sections having capacitance in the series arm of 0 5× 10^-6 farad and in shunt arm an inductance of 0-4× 10^-3 henry. Determine the cut-off frequency.*

Solution:

$$2C = 0.5\times10^{-6} \text{ F} \quad \text{so that} \quad C = 0.25\times10^{-6} \text{ F}$$

and $$L = 0.4\times10^{-3} \text{ H.}$$

The cut-off frequency of a high-pass filter is given by

$$f_c = \frac{1}{4\pi\sqrt{(LC)}}$$

$$= \frac{1}{4\times3.14\sqrt{\left[\left(0.4\times10^{-3}\right)\left(0.25\times10^{-6}\right)\right]}}$$

$$= 8 \times 10^{3} \ c/s = 8 \text{ kc/s.}$$

14

Fourier's Theorem

Periodic Functions

Any function which repeats itself regularly over a given interval of time or space is called a 'periodic' function. Mathematically, a function $f(t)$ is said to be periodic if for all t, there is some positive constant T such that

$$f(t+T) = f(t). \qquad ...(i)$$

T is called the 'period' of $f(f)$.

If r is any integer, then from eq. (i) it follows that

$$f(t+rT) = f(t)$$

for all t. Hence $2T, 3T, 4T, ...$ are also periods of $f(t)$.

For example, $\sin t$ and $\cos t$ are periodic functions having a period 2π. Similarly, the period of $\sin rt$ or $\cos rt$ is $2\pi/r$, while the period of $\tan t$ is π.

Even and Odd Functions: A function $f(t)$ is an even function if

$$f(-t) = f(t). \qquad \text{even}$$

For example t^2, t^4, $\cos t$, $t \sin t$ are even functions.

A function $f(t)$ is an odd function if

$$f(-t) = -f(t). \qquad \text{odd}$$

For example, t, t^3, $\sin t$, $f^2 \sin t$ are odd functions.

Significance of Fourier's Theorem

Fourier's theorem is an important mathematical theorem by which any complex periodic motion can be analysed into a number of simple harmonic components.

It states that 'any finite, continuous, single-valued periodic Junction can be expressed as a summation of simple harmonic terms having frequencies which are multiples of that of the given function'.

Mathematically, the Fourier's theorem can be expressed as:

$$y=f(t)\ A_0+\sum_{r=1}^{\infty} (A_r \cos r\omega t+ B_r \text{Sin } r\omega t)$$

$$=A_0+A_1 \cos \omega t+A_2 \cos 2\omega t +......+A_r \cos r\omega t+......$$

$$+B_1 \sin \omega t+B_2 \sin 2\omega t +......+B_r \sin r\omega t+...... \qquad \text{(i)}$$

where the function $y=f(t)$ is the displacement of a complex periodic motion of angular frequency ω. Thus the complex motion is a sum of cosine and sine components of amplitudes $A_1, A_2, ...; B_1, B_2, ...$ and frequencies which are multiples of ω. A_0 is a constant representing the displacement of the axis of motion from the time-axis.

Limitations :

The Function Should be Finite: This means that the displacement should always have finite values and should never be infinite at any time.

The Function Should be Single-valued: This means that the displacement should have only one value at a given instant t.

The Function Should be Continuous or may just have a finite number of jumps within its time-interval.

These conditions are satisfied in the case of mechanical waves (as sound waves) because no vibrating particle of air or any other medium can show any discontinuity, or have two different displacements at the same time, and also it cannot have an infinite displacement.

Use of F. Theorem in the Analysis of Complex Periodic Motion: Evaluation of Fourier Coefficients: In order to use Fourier theorem for analysing a complex periodic motion or wave-form, we must evaluate the Fourier coefficients A_o, A_r, and B_r.

Evaluation of A_0: Let us multiply eq. (i) by dt and integrate from 0 to T, where T is the period of the function. Then we have

$$\int_0^T y\, dt = A_0 \int_0^T dt + A_1 \int_0^T \cos \omega t\, dt + \ldots + A_r \int_0^T \cos r\omega t\, dt + \ldots..$$

$$+ B_1 \int_0^T \sin \omega t\, dt + \ldots + B_r \int_0^T \sin r\omega t\, dt + \ldots..$$

$= A_o\, T$, all the other integrals being zero.

$$\therefore \boxed{A_0 = \frac{1}{T}\int_0^T y\, dt.} \qquad \ldots\text{(ii)}$$

Physical Significance of A_o: A_0 is the average of the function y over the interval 0 to T. It is, therefore, the steady or 'd.c' level on which the alternating sine and cosine components of the series are superimposed, and the constant A_o can be varied by moving the function with respect to the time-axis. When a periodic function is symmetric about the time-axis, its average value *i.e.* its steady level (A_o) is zero.

Evaluation of A_r: Let us multiply eq. (i) by cos $r\omega t\, dt$ and integrate from 0 to T. Then we have

$$\int_0^T y \cos r\omega t\, dt = A_0 \int_0^T \cos r\omega t\, dt + A_1 \int_0^T \cos \omega t \cos r\omega t\, dt + ...$$

$$+A_r \int_0^T \cos^2 r\omega t\, dt + ...+B_1 \int_0^T \sin \omega t \cos r\omega t\, dt + ...$$

$$+B_r \int_0^T \sin r\omega t \cos r\omega t\, dt + ...$$

$$= A_r \int_0^T \cos^2 r\omega t\, dt, \text{ all the other integrals being zero.}$$

$$= A_r \int_0^T \cos^2 \frac{2\pi rt}{T}\, dt \qquad \left[\because w = \frac{2\pi}{T}\right]$$

$$= A_r \int_0^T \frac{1+\cos(4r\pi t/T)}{2}\, dt$$

$$= A_r \frac{1}{2}\left[t + \frac{\sin(4r\pi/T}{(4r\pi/T)}\right]_0^T$$

$$= A_r \frac{1}{2}\left[T + \frac{\sin(4r\pi)}{4r\pi/T}\right]. \qquad r=1,2....$$

$$= A_r \frac{T}{2}$$

$$\therefore \boxed{A_r = \frac{2}{T}\int_0^T y \cos r\omega t\, dt.} \qquad ...(iii)$$

Evaluation of B_r: Multiplying eq, (i) by sin $rwt\, dt$ and integrating from to T, we shall have

$$\therefore \boxed{B_r = \frac{2}{T}\int_0^T y \cos r\omega t\, dt.} \qquad ...(iv)$$

The interval of integration in (ii), (iii) and (iv) may be replaced by any interval of length T, for example by the interval $-T/2<t<T/2$. We have determined above the coefficients of all the terms on the right-hand-side of eq. (i), which can now be used to analyse a given wave form.

Application in Sound: The quality of a musical note depends upon the harmonics present in it. Hence Fourier's theorem can determine the quality of a note by analysing its oscillation into its harmonic components.

Problems

1. Analyse a square wave with the help of Fourier's theorem.

or

Find the Fourier series components of a complex harmonic motion defined by

$$y = a \text{ when } 0 < t > T/2$$

and $$y = 0 \text{ when } T/2 < t > T.$$

Solution: *Analysis of a Square Wave by Fourier's Theorem:* Let us consider any physical quantity y (such as displacement, pressure or electric current) varying periodically. Here y has a constant value a from time $t=0$ to $t=T/2$ and then remains zero for the remainder of its period T. We may express this function as follows:

$$\boxed{\begin{array}{l} Y = a \text{ from } t=0 \text{ to } t=T/2 \\ Y = 0 \text{ from } t=T/2 \text{ to } t=T, \end{array}}$$

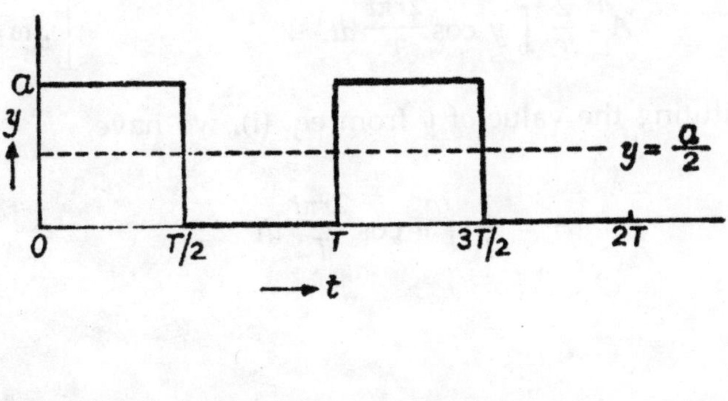

when the time-axis of coordinates is taken through the lowest point of the displacement curve.

The Fourier's theorem can be expressed as

$$y = f(t) = A_O + A_1 \cos \omega t + ...+ A_r \cos r\omega t + ...$$

$$+ B_1 \sin \omega t + ...+ B_r \sin r\omega t + ... \quad ...\text{(ii)}$$

where $A_o = \frac{1}{T}\int_0^T y\,dt$, $A_r = \frac{2}{T}\int_0^T y \cos r\omega t\,dt$

and $B_r = \frac{2}{T}\int_0^T y \sin r\omega t\,dt.$

Let us obtain the values of A_o, A_r and B_r for the given function:

$$A_o = \frac{1}{T}\,\frac{4a}{\pi}\left[\sin \omega t + \frac{1}{3}\sin 3\omega t + \frac{1}{5}\sin 5\omega t + ...\right] dt.$$

Substituting the value of y from eq. (i), we get

$$A_o = \frac{1}{T}\int_0^{T/2} a\,dt\,,$$

i.e. the axis of the displacement curve is the line $y = \frac{a}{2}$.

For A_r, we have

$$A_r = \frac{1}{T}\int_0^T y \cos r\omega t\,dt$$

$$A_r = \frac{2}{T}\int_0^T y \cos \frac{2r\pi t}{T}\,dt. \qquad \left[\because \omega = \frac{2\pi}{T}\right]$$

Substituting the value of y from eq. (i), we have

$$A_r = \frac{1}{T}\int_0^{T/2} a \cos \frac{2r\pi t}{T}\,dt$$

$$= \frac{2a}{T}\ \frac{T}{2r\pi}\left[\sin\frac{2r\pi t}{T}\right]_0^{T/2}$$

$$= \frac{a}{r\pi}[\sin r\pi - \sin 0]$$

$$= 0.$$

Hence all the cosine terms in eq. (ii) are zero.

Similarly, for B_r, we have

$$B_r = \frac{2}{T}\int_0^T y \sin r\omega t\ dt$$

$$= \frac{2}{T}\int_0^T y \sin\frac{2r\pi t}{T}\ dt$$

Substituting the value of y from eq. (i), we have

$$B_r = \frac{2}{T}\int_0^{T/2} a \sin\frac{2r\pi t}{T}\ dt$$

$$= \frac{2a}{T}\ \frac{T}{2r\pi}\left[-\cos\frac{2r\pi t}{T}\right]_0^{T/2}$$

$$= \frac{a}{r\pi}[-\cos r\pi + \cos 0]$$

$$= \frac{a}{r\pi}[-\cos r\pi + 1]$$

When r is even, $\cos r\pi = +1$ and when r is odd, $\cos r\pi = -1$. Thus

$$B_r = \frac{a}{r\pi}\left[-\begin{Bmatrix}1 \text{ for even } r\\ -1 \text{ for odd } r\end{Bmatrix} + 1\right]$$

$$= \frac{a}{r\pi}[-(-1)^r + 1] = \frac{a}{r\pi}\left[1 + (-1)^{r+1}\right]$$

Putting $r = 1, 2, 3, \ldots\ldots$, we get

$$B_1 = \frac{2a}{\pi},\ B_2 = 0,\ B_3 = \frac{2a}{3\pi},\ B_4 = 0,\ B_5 = \frac{2a}{5\pi}, \ldots\ldots$$

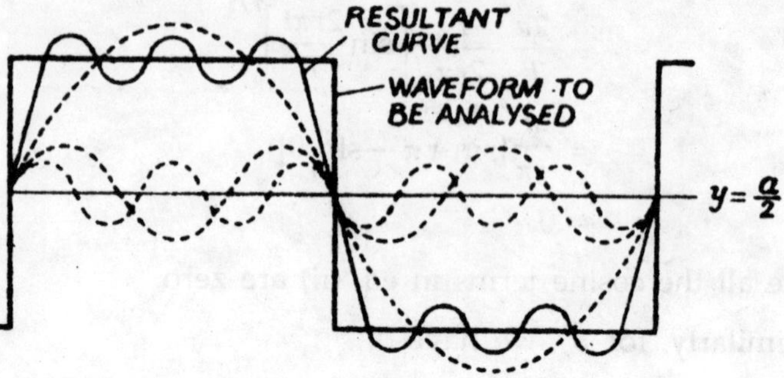

(a)

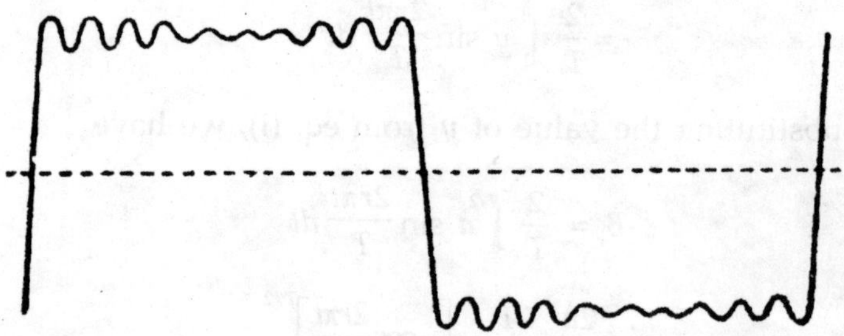

(b)

Therefore, putting the values of A_o, A_r and B_r, eq.(ii) becomes

$$y=\frac{a}{2}+\frac{2a}{\pi}\left[\sin\omega t+\frac{1}{3}\sin 3\omega t+\frac{1}{5}\sin 5\omega t+\ldots\ldots\ldots\right]$$

$$or\ y=\frac{a}{2}+\frac{2a}{\pi}\left[\sin\frac{2\pi t}{T}+\tfrac{1}{3}\sin\frac{6\pi t}{T}+\tfrac{1}{5}\sin\frac{10\pi t}{T}+\ldots\ldots\ldots\right]$$

Thus the given complex motion has the axis $y=a/2$ and the various components as under:

$$\frac{2a}{\pi}\sin\frac{2\pi t}{T}$$

$$\frac{2a}{3\pi}\sin\frac{6\pi t}{T}$$

$$\frac{2a}{5\pi}\sin\frac{10\pi t}{T}, \text{ and so on,}$$

having frequencies in the ratio 1:3:5, etc. and amplitudes in the ratio $1:\frac{1}{3}:\frac{1}{5}$, etc.

In figure above the addition of the successive terms is indicated graphically. The first three terms of the series are shown independently and also the resultant curve. It is seen that even in this case the resultant curve roughly resembles the curve to be analysed. By the addition of 15 terms of the series the curve is almost perfectly reproduced.

2. *Analyse the following curve using Fourier's theorem.*

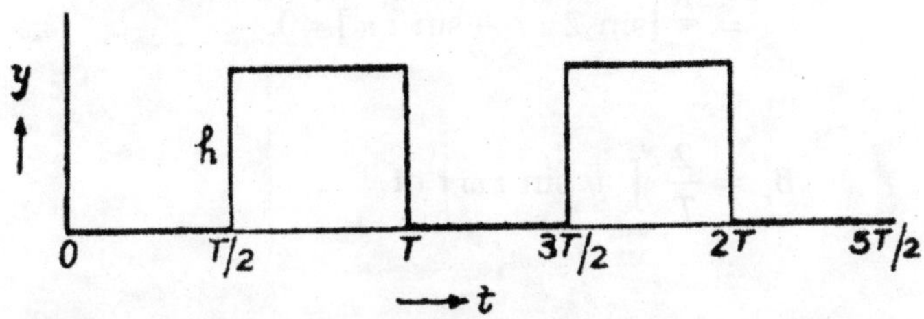

Solution: *Fourier's Analysis of a Square Wave:* The given curve represents a periodic function $y=f(t)$, whose period is T, and which has the value zero from 0 to $T/2$ and the value h from $T/2$ to T, and zero once more, and so on. Mathematically,

$$\boxed{\begin{aligned} &Y=0 \text{ from } t=0 \text{ to } t=T/2 \\ \text{and } &Y=h \text{ from } t=T/2 \text{ to } t=T. \end{aligned}} \quad \text{...(i)}$$

The Fourier series of the function y is

$$y = f(t) = A_0 + A_1 \cos \omega t + ... + A_r \cos r\omega t + ...$$
$$+ B_1 \sin \omega t + ... + B_r \sin r\omega t + ... \quad \text{...(ii)}$$

Let us evaluate the coefficients A_0, A_r and B_r.

$$A_0 = \frac{1}{T}\int_0^T y\, dt$$

$$= \frac{1}{T}\int_{T/2}^{T} h\,dt = \frac{h}{2}$$

$$A_r = \frac{2}{T}\int_{0}^{T} y\cos r\omega t\,dt$$

$$= \frac{2}{T}\int_{0}^{T} y\cos\frac{2r\pi t}{T}\,dt$$

$$= \frac{2}{T}\int_{T/2}^{T} h\cos\frac{2r\pi t}{T}\,dt = \frac{2h}{T}\frac{T}{2r\pi}\left[\sin\frac{2r\pi t}{T}\right]_{T/2}^{T}$$

$$= \frac{h}{r\pi}[\sin 2\pi r - \sin r\pi] = 0.$$

$$B_r = \frac{2}{T}\int_{0}^{T} y\sin r\omega t\,dt$$

$$= \frac{2}{T}\int_{T/2}^{T} h\sin\frac{2r\pi t}{T}\,dt = \frac{2h}{T}\frac{T}{2r\pi}\left[-\cos\frac{2r\pi t}{T}\right]_{T/2}^{T}$$

$$= \frac{h}{r\pi}[-\cos 2\pi r + \cos r\pi] = \frac{h}{r\pi}[-1 + \cos r\pi]$$

$$= \frac{h}{r\pi}\left[-1 + \begin{Bmatrix} 1 \text{ for even } r \\ -1 \text{ for odd } r \end{Bmatrix}\right]$$

$$= \frac{h}{r\pi}[-1 + (-1)^r]$$

Putting $r = 1, 2, 3$

$$B_1 = -\frac{2h}{\pi},\ B_2 = \theta,\ B_3 = -\frac{2h}{3\pi},\ B_4 = ,0\ B_5 = -\frac{2h}{5\pi}, \ldots\ldots$$

Putting the values of A_0 , A_r and B_r in eq. (ii), we get

$$y = \frac{h}{2} - \frac{2h}{\pi} \sin \omega t - \frac{2h}{3\pi} \sin 3\omega t - \frac{2h}{5\pi} \sin 5\omega t......$$

or $$y = \frac{h}{2} - \frac{2h}{\pi}\left[\sin\frac{2\pi t}{T} + \frac{1}{3}\sin\frac{6\pi t}{T} + \frac{1}{5}\sin\frac{10\pi t}{T} + ...\right].$$

3. Obtain the Fourier series for a square wave defined by the continuous harmonic function

$$y = a \text{ when } 0 \le t \le T/2$$

and $$y = -a \text{ when } T/2 \le t \le T.$$

Solution: *Analysis of a Harmonic Function (Square Wave) by Fourier's Theorem:* The given harmonic function

and $$\boxed{\begin{array}{l} Y = a \text{ from } t = 0 \quad \text{to } t = T/2 \\ Y = a \text{ from } t = T/2 \text{ to } t = T, \end{array}} \qquad ...(i)$$

represents a square wave as shown in Figure.

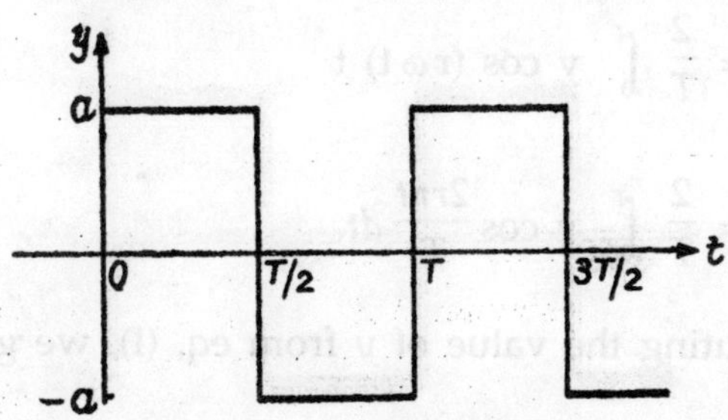

The Fourier series of a harmonic function is

$$y = f(t) = A_0 + A_1 \cos \omega t + ... + A_r \cos r\omega t + ...$$

$$+ B_1 \sin \omega t + ... + B_r \sin r\omega t + ..., \qquad ...(ii)$$

where $$Ao = \frac{1}{T}\int_0^T y\, dt,\; A_r = \frac{2}{T}\int_0^T y \cos r\omega t\, dt$$

and $$Br = \frac{2}{T}\int_0^T y \sin r\omega t \, dt.$$

Let us obtain the values of A_0. A_t and B_r, for the given function:

$$A_0 = \frac{1}{T}\int_0^T y \, dt.$$

Substituting the value of y from eq. (i), we get

$$A_0 = \frac{1}{T}\int_0^{T/2} a \, dt - \frac{I}{T}\int_{T/2}^{T} a \, dt$$

$$= \frac{a}{T}[t]_0^{T/2} - \frac{a}{T}[t]_{T/2}^{T}$$

$$= 0.$$

Thus the axis of the displacement curve is the x-axis (y=0).

For A_r, we have

$$A_r = \frac{2}{T}\int_0^T y \cos (r\omega t) \, t$$

$$= \frac{2}{T}\int_0^T y \cos\frac{2r\pi t}{T} dt. \qquad \left[\because \omega \frac{2\pi}{T}\right]$$

Substituting the value of y from eq. (i), we get

$$A_r = \frac{2}{T}\int_0^{T/2} a \cos\frac{2r\pi t}{T} dt - \frac{2}{T}\int_{T/2}^{T} a \cos\frac{2r\pi t}{T} dt$$

$$= \frac{2a}{T}\frac{T}{2r\pi}\left[\sin\frac{2r\pi t}{T}\right]_0^{T/2} - \frac{2a}{T}\left[\sin\frac{2r\pi t}{T}\right]_{T/2}^{T}$$

$$= \frac{a}{r\pi}[\sin r\pi - \sin 0]\frac{a}{r\pi}[\sin 2r\pi - \sin r\pi]$$

$$= \frac{a}{r\pi}[2\sin r\pi - \sin 2r\pi] = 0.$$

Hence all the cosine term in eq. (ii) are zero.

For B_r, we have

$$B_r = \frac{2}{T}\int_0^T y \sin(r\omega t)\, dt$$

$$= \frac{2}{T}\int_0^T y \sin\frac{2r\pi t}{T}\, dt$$

$$= \frac{2}{T}\int_0^{T/2} a \sin\frac{2r\pi t}{T}\, dt - \frac{2}{T}\int_{T/2}^{T} a \sin\frac{2r\pi t}{T}\, dt$$

$$\frac{2a}{T}\frac{T}{2r\pi}\left[-\cos\frac{2r\pi t}{T}\right]_0^{T/2} - \frac{2}{T}\frac{T}{2r\pi}\left[-\cos\frac{2r\pi t}{T}\right]_{T/2}^{T}$$

$$= \frac{a}{r\pi}[-\cos r\pi + \cos 0] - \frac{a}{r\pi}[-\cos 2r\pi + \cos r\pi)$$

$$= \frac{a}{r\pi}[-\cos r\pi + 1 + \cos 2r\pi - \cos r\pi]$$

$$= \frac{a}{r\pi}[2 - 2\cos r\pi].$$

$$= \frac{2a}{r\pi}[1 - \cos r\pi].$$

But $\cos r\pi = (-1)^r$, where $r = 1, 2, 3, \ldots\ldots\ldots.$

$$\therefore\ B_r = \frac{a}{r\pi}[1-(-1)^r].$$

Putting $r = 1, 2, 3, \ldots\ldots\ldots.$, we get

$$B_1 = \frac{4a}{\pi},\ B_2 = 0,\ B_3 = \frac{4a}{3\pi},\ B_4 = 0,\ B_5 = \frac{4a}{5\pi},\ldots\ldots$$

Putting the values of A_o, A_r and B_r, in eq. (ii), we get

$$y = \frac{4a}{\pi}\left[\sin\omega t + \frac{1}{3}\sin 3\omega t + \frac{1}{5}\sin 5\omega t + ...\right]$$

or $$y = \frac{4a}{\pi}\left[\sin\frac{2\pi t}{T} + \frac{1}{3}\sin\frac{6\pi t}{T} + \frac{1}{5}\sin\frac{10\pi t}{T} + ...\right]$$

Thus the given function can be written as a sum of simple harmonic components having frequencies in the ratio 1:3:5........ and amplitudes in the ratio $1 : \frac{1}{3} : \frac{1}{5} :$..... .

If we take the first three terms of the above series representing the square wave and add them together, the result is figure given below.

The first harmonic has the frequency of the square-wave and the higher frequencies build up the squareness of the wave. The highest frequencies are responsible for the sharpness of the vertical sides of the wave.

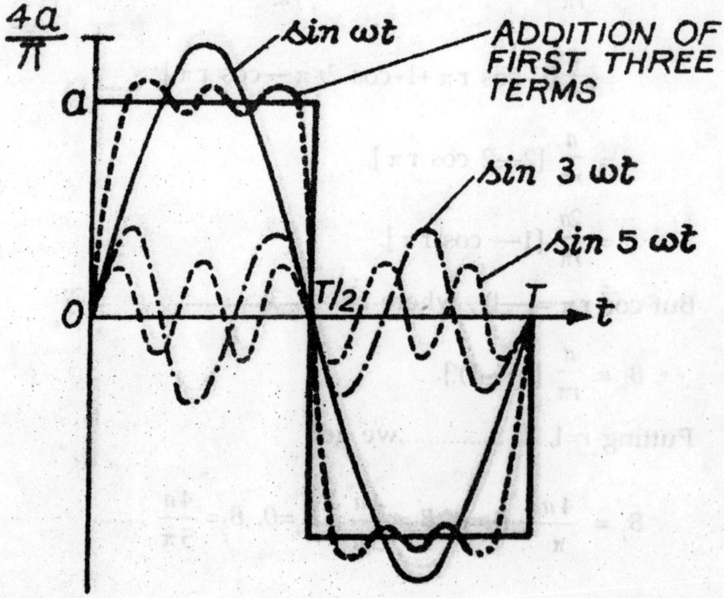

4. Analyse the following curve using Fourier's theorem.

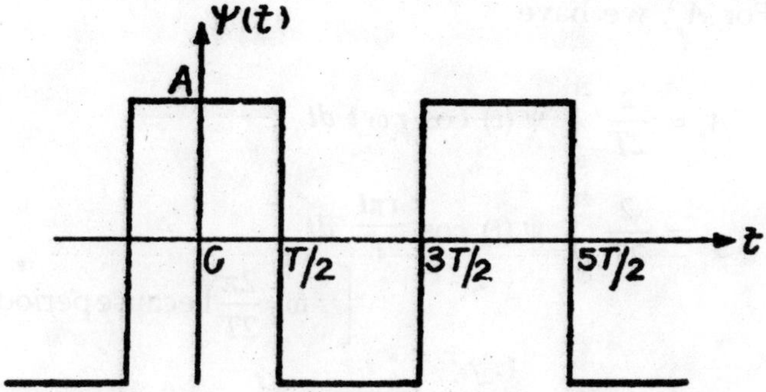

Solution: *Fourier's Analysis of the given Curve:* The given curve represents a periodic function: $\psi\ (t)$, of period $2T$ which has the value A from — $T/2$ to $T/2$ and value A from $T/2$ to $3T/2$, that is

$$\boxed{\begin{aligned}\psi\ (t) &= A \text{from} t = T/2 to t = T/2 \\ \psi(t) &= A \text{from} t = T/2 to t = 3T/2\end{aligned}}$$

and ...(i)

The Fourier series of the function is

$$\psi\ (t) = A_0 + A_1 \cos\ \omega\, t + \ldots\ldots + A_r \cos\ r\,\omega\, t + \ldots.$$

$$+ B_1 \sin\ \omega\, t + \ldots\ldots + B_r \sin\ r\,\omega\, t + \ldots. \qquad \ldots\text{(ii)}$$

Let us evaluate coefficients A_o, A_r and B_r.

$$A_0 = \frac{1}{2T}\int_0^{2T} \psi\ (t)\ dt$$

$$= \frac{1}{2T}\int_{-T/2}^{T/2} A\ dt + \frac{1}{2T}\int_{T/2}^{3T/2} -A\ dt$$

$$= \frac{A}{2T}\ [t]_{-T/2}^{T/2} - \frac{A}{2T}\,[\,t\,]_{T/2}^{3T2}$$

$$= \frac{A}{2T}\left[\frac{T}{2}+\frac{T}{2}\right] - \frac{A}{2T}\left[\frac{3T}{2}+\frac{T}{2}\right]$$

For A_r, we have

$$A_r = \frac{2}{2T} \int_0^{2T} \psi(t) \cos r\omega t \, dt$$

$$= \frac{2}{2T} \int_0^{2T} \psi(t) \cos \frac{r\pi t}{T} \, dt$$

$$\left[\because \omega = \frac{2\pi}{2T} \text{ because period is } 2T\right]$$

$$= \frac{1}{T} \int_{-T/2}^{T/2} A \cos \frac{r\pi t}{T} \, dt - \frac{1}{T} \int_{T/2}^{3T/2} A \cos \frac{r\pi t}{T} \, dt$$

$$= \frac{A}{T} \frac{T}{r\pi} \left[\sin \frac{r\pi t}{T}\right]_{-T/2}^{T/2} - \frac{A}{T} \frac{T}{r\pi} \left[\sin \frac{r\pi t}{T}\right]_{T/2}^{3T/2}$$

$$= \frac{A}{r\pi} \left[\sin \frac{r\pi}{2} + \sin \frac{r\pi}{2}\right] - \frac{A}{r\pi} \left[\sin \frac{3r\pi}{2} + \sin \frac{r\pi}{2}\right]$$

$$= \frac{A}{r\pi} \left[3 \sin \frac{r\pi}{2} - \sin \frac{3r\pi}{2}\right]$$

When r is even, both the sine terms become zero. When r is odd (1, 3, 5, ...), then

$$\sin \frac{r\pi}{2} = 1, -1, 1, \ldots\ldots$$

and $$\sin \frac{3r\pi}{2} = -1, -1, 1, \ldots\ldots$$

Thus putting $r=1$, 2, 3, 4, 5, ... we have

$$A_1 = \frac{4A}{\pi} \quad A_2 = 0, \; A_3 = -\frac{4A}{3\pi}, \; A_4 = 0, \; A_5 = \frac{4A}{5\pi}, \ldots\ldots$$

Proceeding as above, we can see that

$B_r = 0$ for all values of r.

Hence the Fourier expansion of the given function is

$$\psi(t) = \frac{4A}{\pi}\left[\cos \omega t - \frac{1}{3}\cos 3\omega t + \frac{1}{5}\cos 5\omega t - \frac{1}{7}\cos 7\omega t + ...\right]$$

But here $\omega = \frac{2\pi}{2T} = \frac{\pi}{T}$

$$\therefore \psi(t) = \frac{4A}{\pi}\left[\cos\frac{\pi t}{T} - \frac{1}{3}\cos\frac{3\pi t}{T} + \frac{1}{5}\cos\frac{5\pi t}{T} .....\right]$$

5. *Analyse the following saw-tooth curve into its components.*

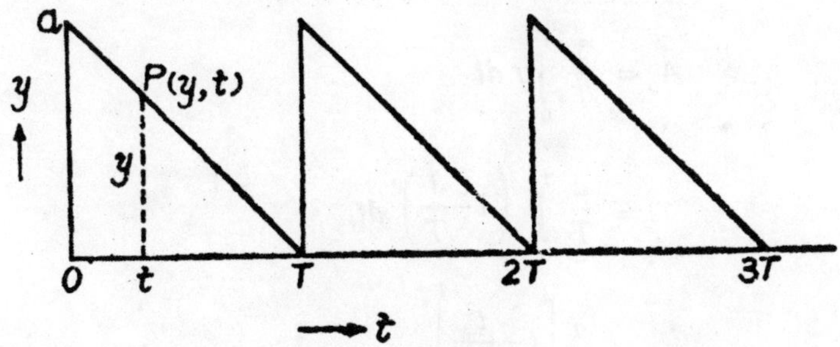

or

Obtain the Fourier series for the saw-tooth wave:

$y = f(t) = a\left(1 - \frac{t}{T}\right)$, for $0 < t < T$.

or

Apply Fourier's theorem to analyse a saw-tooth wave into its hart the components.

Solution: *Analysis of a Saw-tooth Wave by Fourier's Theorem:* The curve represents a saw-tooth wave, it can be expressed mathematically as*

$$y = f(t) = a\left(1 - \frac{t}{T}\right), \text{ from } t=0 \text{ to } t=T. \qquad ...(i)$$

The Fourier's series is

$y = f(t) = A_0 + A_1 \cos \omega t + ...+ A_r \cos r\omega t + ...$

$$+B_1 \sin \omega t + ...+B_r \sin r\omega t+....... \quad ...(ii)$$

where $A_0 = \frac{1}{T}\int_0^T y\, dt,\ A_r = \frac{2}{T}\int_0^T y \cos r\omega t\, dt$

and $B_r = \frac{2}{T}\int_0^T y \sin r\omega t\, dt.$

Let us obtain the values of the coefficients A_0, A_r and B_r of the Fourier series:

$$A_0 = \frac{1}{T}\int_0^T y\, dt$$

$$= \frac{a}{T}\int_0^T \left(1-\frac{t}{T}\right) dt,$$

$$= \frac{a}{T}\left[t-\frac{t^2}{2T}\right]_0^T$$

$$= \frac{a}{T}\left[T-\frac{T^2}{2T}\right]$$

$$= \frac{a}{2},$$

which gives the ordinate of the axis of the curve.

For A_r, we have

$$A_r = \frac{2}{T}\int_0^T y \cos r\omega t\, dt$$

$$= \frac{2}{T}\int_0^T y \cos \frac{2r\pi t}{T}\, dt. \qquad \left[\because \omega = \frac{2\pi}{T}\right]$$

Substituting the value of y from eq. (i), we have

$$A_r = \frac{2}{T}\int_0^T a\left(1-\frac{t}{T}\right)\cos\frac{2r\pi t}{T}\,dt$$

$$= \frac{2a}{T}\int_0^T \cos\frac{2r\pi t}{T}\,dt - \frac{2a}{T}\int_0^T t\cos\frac{2r\pi t}{T}\,dt$$

$$= -\frac{2a}{T^2}\int_0^T t\cos\frac{2r\pi t}{T}\,dt \left[\because \int_0^T \cos\frac{2\pi t}{T}dt = o\right]$$

$$= -\frac{2a}{T^2}\left[t\frac{\sin(2\pi/T)}{2r\pi/T} - \int\frac{\sin(2\pi t/T)}{(2r\pi/T)^2}\right]_0^T \text{ (by parts)}$$

$$= -\frac{2a}{T^2}\left[t\frac{\sin(2\pi/T)}{2r\pi/T} + \frac{\sin(2\pi t/T)}{(2r\pi/T)^2}\right]_0^T$$

$$= -\frac{2a}{T^2}\left[T\frac{\sin 2r\pi}{2r\pi/T} + \frac{\cos 2r\pi}{(2r\pi/T)^2} - 0 - \frac{\cos 0}{(2r\pi/T)^2}\right]$$

$$= -\frac{2a}{T^2}\left[\frac{1}{(2r\pi/T)^2} - \frac{1}{(2r\pi/T)^2}\right]$$

$$= 0. \qquad [\because \sin 2r\pi = 0 \text{ and } \cos 2r\pi = 1]$$

For B_r, we have

$$B_r = \frac{2}{T}\int_0^T y\sin r\omega t\,dt$$

$$= \frac{2}{T}\int_0^T y\sin\frac{2r\pi t}{T}\,dt. \qquad \left[\because \omega = \frac{2\pi}{T}\right]$$

Substituting the value of y from eq. (i), we get

$$B_r = \frac{2}{T}\int_0^T a\left(1-\frac{t}{T}\right)\sin\frac{2r\pi t}{T}\,dt$$

$$= \frac{2a}{T}\int_0^T \sin\frac{2r\pi t}{T}\,dt - \frac{2a}{T^2}\int_0^T t\sin\frac{2r\pi t}{T}\,dt$$

$$= -\frac{2a}{T^2}\int_0^T t\sin\frac{2r\pi t}{T}\,dt \quad \left[\because \int_0^T \sin\frac{2\pi t}{T}dt = o\right]$$

$$= -\frac{2a}{T^2}\left[t\frac{-\cos(2\pi/T)}{2r\pi/T} - \int \frac{-\cos(2\pi t/T)}{2r\pi/T}\right]_0^T \text{ (by parts)}$$

$$= \frac{2a}{T^2}\left[t\frac{\cos(2\pi/T)}{2r\pi/T} - \frac{\sin(2\pi t/T)}{(2r\pi/T)^2}\right]_0^T$$

$$= \frac{2a}{T^2}\left[T\frac{\cos 2r\pi}{2r\pi/T} - \frac{\sin 2r\pi}{(2r\pi/T)^2} - 0 + \frac{\sin 0}{(2r\pi/T)^2}\right]$$

$$= -\frac{2a}{T^2}\left[T\frac{1}{2r\pi/T}\right] \quad [\because \cos 2r\pi = 1 \text{ and } \sin 2r\pi = 0]$$

$$= \frac{a}{r\pi}$$

Thus $B_1 = \frac{a}{\pi}$, $B_2 = \frac{a}{2\pi}$, $B_3 = \frac{a}{3\pi}$,...........

Substituting the values of A_o, A_r and B_r, in eq. (ii), we have

$$y = \frac{a}{2} + \frac{a}{\pi}(\sin \omega t + \frac{1}{2}\sin 2\omega t + \frac{1}{3}\sin 3\omega t + \ldots\ldots)$$

or $$y = \frac{a}{2} + \frac{a}{\pi}(\sin\frac{2\pi t}{T} + \frac{1}{2}\sin\frac{4\pi t}{T} + \frac{1}{3}\sin\frac{6\pi t}{T} + \ldots\ldots)$$

Thus the given periodic motion has the axis $y = \frac{a}{2}$ above the time-axis, and the various components as under

$$\frac{a}{\pi}\sin\frac{2\pi t}{T},$$

$$\frac{a}{2\pi}\ \sin\frac{4\pi t}{T},$$

$$\frac{a}{3\pi}\ \sin\frac{6\pi t}{T},$$

and soon, having frequencies in the ratio 1:2: 3...and amplitudes in the ratio $1:\frac{1}{2}:\frac{1}{3}:......$.

The addition of successive terms of this series is indicated graphically in following figure. It is seen that greater the number of terms

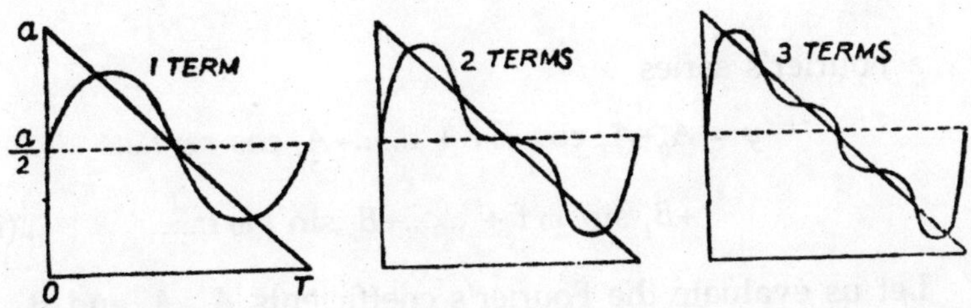

used, the closer the resemblance between the resultant curve and the curve under analysis.

6. Find the Fourier's series for the saw-tooth function given by

$$y = \frac{a}{T}t \qquad \textit{for } 0 < t < T.$$

$$\textit{Ans. } y = \frac{a}{2} - \frac{a}{\pi}\left(\sin\frac{2\pi t}{T} + \tfrac{1}{2}\sin\frac{4\pi t}{T} + \tfrac{1}{3}\ \sin\frac{6\pi t}{T} + ...\right.$$

7. Use Fourier's theorem to analyse the function

$$y = a\left(1 - \frac{2t}{T}\right) \quad \text{for t= 0 to t = T.}$$

Solution: *Analysis of Saw-tooth Function:* The given function

$$y = a\left(1 - \frac{2t}{T}\right), \text{ for } 0 < t < T. \qquad ...(i)$$

represents a saw-tooth wave. It can be written as

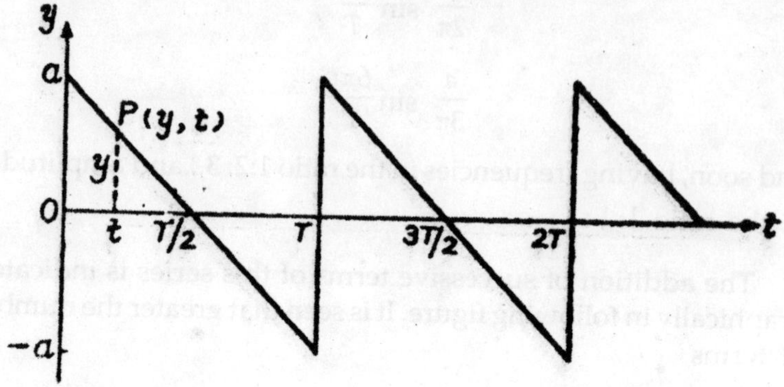

a Fourier's series

$$y = A_0 + A_1 \cos \omega t + ...+ A_r \cos r \omega t.....$$

$$+B_1 \sin \omega t + ...+ B_r \sin r \omega t..... \qquad ...(i)$$

Let us evaluate the Fourier's coefficients A_o, A_r and B_r.

$$A_0 = \frac{1}{T} \int_0^T y \, dt$$

$$= \frac{1}{T} \int_0^T a \left(1 - \frac{2t}{T}\right) dt$$

$$= \frac{a}{T} \int_0^T \left[t - \frac{t^2}{T}\right]_0^T$$

we find

$$A_r = 0$$

and $\quad B_r = \dfrac{2a}{r\pi}$

Thus $\quad B_1 = \dfrac{2a}{\pi}, \; B_2 = \dfrac{2a}{2\pi}, \; B_3 = \dfrac{2a}{3\pi}, ...$

Substituting these values in eq. (ii) we get

$$y = \frac{2a}{\pi}(\sin\ \omega\ t + \frac{1}{2}\sin 2\omega\ t + \frac{1}{3}\sin 3\omega\ t + \ldots\ldots)$$

or $$y = \frac{2a}{\pi}(\sin\frac{2\pi t}{T} + \frac{1}{2}\sin\frac{4\pi t}{T} + \frac{1}{3}\sin\frac{6\pi t}{T} + \ldots\ldots)$$

The addition of successive terms of this series is shown graphically in figure given below. It is seen that greater the number of terms used, the closer the resemblance between the resultant curve and the curve under analysis.

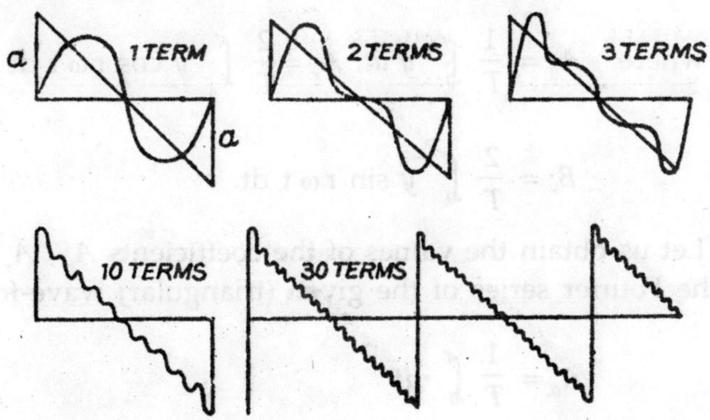

8. Analyse a triangular wave into its simple harmonic components with the help of Fourier's theorem.

Solution: *Analysis of a Triangular Wave:* Let us consider a triangular wave-form, given by

$y = 0$ at $t = 0$

$y = a$ at $t = T/2$

and $y = 0$ at $t = T$,

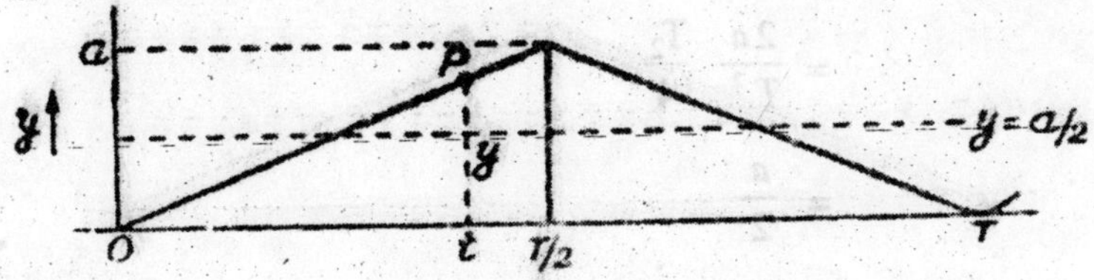

and the curve repeats itself after $t=T$. The displacement at any time/may be written as

$$y = \frac{2at}{T} \quad \text{form} \quad t = 0 \text{ to } t = T/2$$

and

$$y = \frac{2a(T-t)}{T} \quad \text{from} \quad t = o \text{ to } t = T.$$

Now, the Fourier's series is

$$y = f(t) = A_0 + A_1 \cos \omega t + \ldots\ldots + A_r \cos r\omega t + \ldots\ldots$$

$$+ B_1 \sin \omega t + \ldots\ldots + B_r \sin r\omega t + \ldots\ldots$$

where $A_0 = \frac{1}{T}\int_0^T y\, dt,\ A_r = \frac{2}{T}\int_0^T y \cos r\omega t\, dt$

$$B_r = \frac{2}{T}\int_0^T y \sin r\omega t\, dt.$$

Let us obtain the values of the coefficients A_o, A_r and B_r of the Fourier series of the given (triangular) wave-form:

$$A_0 = \frac{1}{T}\int_0^T dt$$

$$= \frac{1}{T}\left[\int_0^{T/2} \frac{2at}{T} dt + \int_{T/2}^{T} \frac{2a(T-t)}{T} dt\right]$$

$$= \frac{1}{T}\left[\frac{2a}{T}\left(\frac{t^2}{2}\right)_0^{T/2} + \frac{2a}{T}\left(Tt - \frac{t^2}{2}\right)_{T/2}^{T}\right]$$

$$= \frac{2a}{T^2}\left[\frac{T^2}{8} + \left(T^2 - \frac{T^2}{2}\right) - \left(\frac{T^2}{2} - \frac{T2}{2}\right)\right]$$

$$= \frac{2a}{T^2}\ \frac{T_2}{4}$$

$$= \frac{a}{2}$$

This is the displacement of the axis of the curve from the axis of the coordinate system. Now,

$$A_r = \frac{2}{T}\int_0^T y \cos r\omega t\, dt$$

$$= \frac{2}{T}\left[\int_0^{T/2} \frac{2at}{T}\cos r\omega t\, dt + \int_{T/2}^{T} \frac{2a(T-t)}{T}\cos r\omega t\, dt\right]$$

$$= \frac{4a}{T^2}\left[\int_0^{T/2} t\cos r\omega t\, dt + \int_{T/2}^{T} T\cos r\omega t\, dt - \int_{T/2}^{T} t\cos r\omega t\, dt\right]$$

On solving these integrals, we obtain

$$A_r = \frac{4a}{T^2}\left[\frac{2}{r^2\omega^2}\{(\cos \pi r)\} - 1\right]$$

$$= \frac{4a}{T^2}\left[\frac{2}{r^2(2\pi/T)^2}\left\{(-1)^r - 1\right\}\right] \qquad [\because \cos \pi r = (-1)^r]$$

$$= \frac{2a}{\pi^2 r^2}[(-1)^r - 1]$$

$$= 0, \text{ if r is even}$$

$$= -\frac{4a}{\pi^2 r^2}, \text{ if r is odd.}$$

Thus, only odd harmonics with cosine terms would appear in the Fourier series.

$$B_r = \frac{2}{T}\int_0^T y \sin r\omega t\, dt$$

$$= \left[\int_0^{T/2} \frac{2at}{T}\sin r\omega t\, dt + \int_{T/2}^{T} \frac{2a(T-t)}{T}\sin r\omega t\, dt\right]$$

$$= \frac{4a}{T^2}\left[\int_0^{T/2} t \sin r\omega t\, dt + \int_{T/2}^{T} T \sin r\omega t\, dt - \int_{T/2}^{T} t \sin r\omega t\, dt\right]$$

$$= 0.$$

Thus all the sine terms in the Fourier series are zero.

Substituting the values of A_o, A_r and B_r in the Fourier series, we obtain

$$y = \frac{a}{2} + \frac{4a}{\pi^2} \cos \omega t - \frac{4a}{3^2\pi^2} \cos 3\,\omega t - \frac{4a}{5^2\pi^2} \cos 5\,\omega t + \ldots\ldots)$$

$$\text{or} \quad = \frac{a}{2} + \frac{4a}{\pi^2} (\cos\ \omega t\ \frac{1}{3^2} \cos 3\,\omega t + \frac{1}{5^2} \cos\ 5\,\omega t \ldots\ldots).$$

Thus the given triangular vibration has the axis $y=a/2$ above the time-axis and the various cosine components of amplitude coefficients,

$$-\frac{4a}{\pi^2}\left(1, \frac{1}{3^2}, \frac{1}{5^2}, \frac{1}{7^2}, \ldots\ldots\right)$$

and frequencies in the ratio 1:3:5:7

The resultant obtained by adding three terms of this series is shown in figure given below. As the series is rapidly convergent, the curve under analysis is closely represented by the sum of only a few terms of the series.

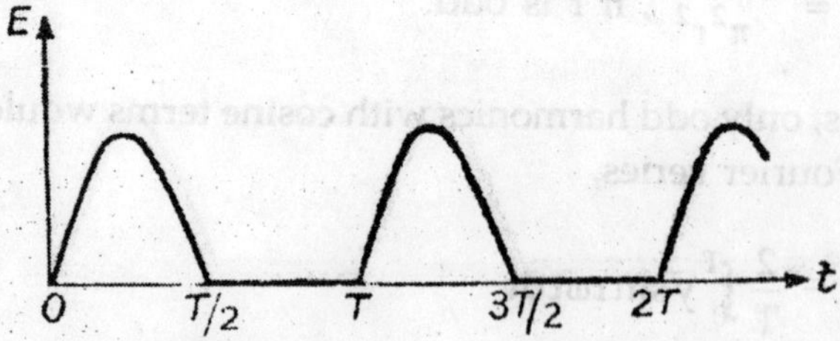

9. Apply Fourier's theorem to analyse the output wave from a half-wave rectifier when the input wave is of the form $E=E_o \sin \omega t$.

Or

Analyse the following curve using Fourier's theorem.

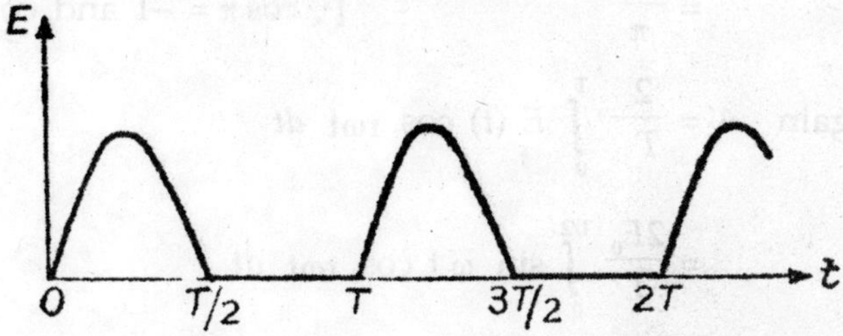

***Solution:** Fourier's Analysis of the Output Wave from a Half-wave Rectifier:* A sinusoidal voltage $E = E_0 \sin \omega t$ is passed through a half-wave rectifier which removes the negative half-cycles of the wave. The output voltage wave is of the form shown in figure above. This may be expressed as

$$E(t) = \begin{cases} E_0 \sin\omega t \text{ from } t = 0 \text{ to } t = T/2 \\ 0 \text{ from } t = T/2 \text{ to } t = T/2 \text{ to} = T. \end{cases} \quad T = \frac{2\pi}{\omega} \quad \text{...(i)}$$

Let us express it as a Fourier series:

$$E(t) = A_o + A_1 \cos \omega t + \text{.........} + A_r \cos r\omega t + \text{...}$$

$$+ B_1 \sin \omega t + \text{...........} + B_r \sin r\omega + \text{....} \quad \text{...(ii)}$$

Let us evaluate the Fourier coefficients:

$$A_0 = \frac{1}{T}\int_0^T E(t)\, dt$$

$$= \frac{1}{T}\int_0^{T/2} E_0 \sin\, dt$$

$$= \frac{1}{T}\int_0^{T/2} E_0 \sin\frac{2\pi t}{T}\, dt$$

$$= -\frac{E_0}{T}\,\frac{1}{2\pi/T}\left[\cos\frac{2\pi t}{T}\right]_0^{T/2}$$

$$= -\frac{E_0}{2\pi}\left[\cos\pi - \cos 0\right]_0^0$$

$$= \frac{E_0}{\pi} \qquad [\because \cos\pi = -1 \text{ and } \cos 0 = 1]$$

Again $A_r = \frac{2}{T}\int_0^T E(t)\cos r\omega t\, dt$

$$= \frac{2E_0}{T}\int_0^{T/2} \sin\omega t \cos r\omega t\, dt$$

$$= \frac{2E_0}{T}\int_0^{T/2} \sin\frac{2\pi t}{T}\cos\frac{2r\pi t}{T}\, dt$$

$$= \frac{2E_0}{T}\int_0^{T/2} \tfrac{1}{2}\left[\sin(1+r)\frac{2\pi t}{T} + \sin(1-r)\frac{2\pi t}{T}\right] dt$$

$$= \frac{E_0}{T}\left[-\frac{\cos(1+r)2\pi t/T}{(1+r)2\pi/T} - \frac{\cos(1-r)2\pi t/T}{(1-r)2\pi/T}\right]_0^{T/2}$$

$$= \frac{E_0}{T}\left[-\frac{\cos(1+r)\pi - \cos}{(1+r)2\pi/T} + \frac{\cos(1-r)\pi - \cos 0}{(1-r)2\pi/T}\right]$$

$$= \frac{E_0}{2\pi}\left[\frac{\cos(1+r)\pi - 1}{1+r} + \frac{\cos(1-r)\pi - 1}{1-r}\right]$$

When r is odd then this is equal to zero, because cos 0=cos 2 π =cos 4 π -1.

When r is even, then we have

$$A_r = \frac{E_0}{2\pi}\left[\frac{2}{1+r} + \frac{2}{1-r}\right] \; [\because \cos\pi = \cos 3\pi = \cos 5\pi = -1]$$

$$= \frac{E_0}{2\pi}\,\frac{2}{(1+r)(1-r)}$$

$$= \frac{2E_0}{\pi}\frac{1}{(1+r)(1-r)};\ r=2,\ 4,\ 6,......$$

$$= \quad -\frac{2E_0}{\pi}\left[\frac{1}{(1)(3)},\frac{1}{(3)(5)},\frac{1}{(5)(7)},....\right].$$

Again $B_r = \frac{2}{T}\int_0^{T/2} E\ (t) \sin\ r\omega t\ dt$

$$= \frac{2E_0}{T}\int_0^{T/2} \sin\ \frac{2\pi t}{T} \sin \frac{2\pi t}{T}\, dt$$

Proceeding as above, we can see that

$$B_r = \frac{E_0}{2} \qquad \text{for } r = 1$$

and $B_r = 0$ for r = 2, 3,4,........

Substituting these values of A_0, A_r and B_r in eq. (ii), we get

$$E\ (t)\ \frac{2E_0}{\pi} + \frac{2E_0}{\pi}\ \sin\ \omega\, t - \frac{2E_0}{\pi}\left[\frac{1}{(1)(3)}\cos\ 2\,\omega\ t + \frac{1}{(3)(5)}\cos\ 4\,\omega\ t + \frac{1}{(5)(7)}\cos\ 6\omega t + ...\right].$$

15

Doppler's Effect

Whenever there is a relative motion between an observer and a distant sounding source, the pitch (frequency) of the sound as heard by the observer is different from its true pitch. The pitch heard appears to be higher than the true pitch if the distance between the source and the observer is decreasing, and vice versa.

When a whistling engine approaches an observer standing on the platform, the pitch of the note heard by him is higher than its true pitch. As the engine passes the observer, there is a sudden drop in pitch, the pitch of the note heard by the observer being now lower than its true pitch. A similar drop is noticed when a fast low-flying air craft passes over head. The change in pitch is also experienced when the source is stationary and the observer is moving. This phenomenon is called "Doppler's effect". Its explanation is as follows:

Motion of the Source: Let S be a sound source emitting a note of true frequency n, and O a stationary observer. Let v be the velocity of the sound waves.

If the source S is also stationary, then the n waves emitted by it in an interval of one second will occupy a distance $SO=v$. The first wave emitted at the beginning of the interval has arrived at O when the last (nth) wave is just being emitted from a at the end of the interval. The wavelength is thus

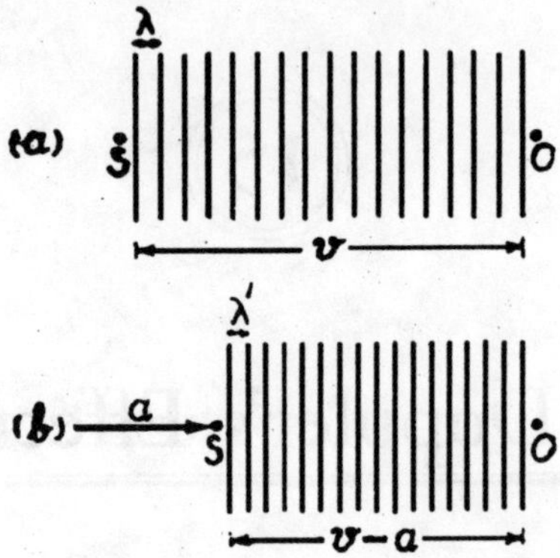

$$\lambda = \frac{v}{n}$$

If now S is moving towards O with a velocity a. the first wave emit at the beginning of an interval of one second will still have arrived at O, but the nth wave at the end of the interval will be emitted when S has moved through a distance a towards O. Thus n waves are crowded into a distance v–a. Consequently, the wavelength will be decreased to λ given by

$$\lambda = \frac{v-a}{n}$$

The wavelength being decreased, the frequency will be increased. If the apparent frequency heard by the observer is n′, we have

$$n' = \frac{v}{\lambda'} = \frac{v}{(v-a)/n}$$

or

$$n' = n\left(\frac{v}{v-a}\right)$$

If the source is moving away from the observer, the n^1 waves emitted by it in one second will occupy a distance *(v+a)*; and the wavelength will be increased to *(v+a)/n*. The apparent frequency will then be given by

$$n' = n\left(\frac{v}{v+a}\right)$$

Motion of the Observer: When the observer O is at rest, *n* waves (of wavelength A) pass him in one second. But if he is moving towards the source S with a velocity *b*, additional number of waves equal to that contained in a length *b* will pass him per second. Therefore, the total number of waves received by him per second, that is, the apparent frequency of the source will be given by

$$n' = n + \frac{b}{\lambda}$$

$$= n + \frac{b}{v/n} \qquad \left[\because \lambda = \frac{v}{n}\right]$$

$$= n\left(1 + \frac{b}{v}\right)$$

$$n' = n\left(\frac{v+a}{v}\right). \qquad ...(iii)$$

If the observer is moving away from *the source,* he will receive in one second n waves minus the number of waves contained in the distance *b*. Therefore, the apparent frequency will be given by

$$n' = n - \frac{b}{\lambda}$$

$$= n - \frac{b}{v/n} \qquad \left[\because \lambda = \frac{v}{n}\right]$$

$$= n\left(1 - \frac{b}{v}\right)$$

$$n' = n\left(\frac{v-a}{v}\right). \qquad \text{...(iv)}$$

Motion of both the Source and the Observer: Suppose, both the source and the observer are moving with velocities *a* and *b* respectively in the same direction as the sound.

If the observer be at rest, the apparent frequency, say n_1, is by eq. (i), given by

$$n_1 = n\left(\frac{v}{v-a}\right).$$

If the observer is also moving away from the source, the apparent frequency is further altered to the value n', which, by eq. (iv), is given by

$$n' = n_1\left(\frac{v-a}{v}\right).$$

Substituting the value of n_1 from above, we get

$$n' = n\left(\frac{v-a}{v-a}\right). \qquad \text{...(v)}$$

Effect of Wind: If the wind is blowing with a velocity *w* in the same direction as the sound, then the sound waves will cover a total distance *(v+w)* in one second. Hence, replacing

v by *(v+w) in* eq. (v). the apparent frequency will then by given by

$$n' = n\left(\frac{v+w-b}{v+w-a}\right).$$

Practical Application

An interesting application of the Doppler effect is to detect moving submarine under water and the aeroplanes in air and to estimate their velocities. Waves of ultrasonic frequency are transmitted from the ship and their echoes from the submarine and the sea bed are received. A difference between the frequencies of the two reflected waves indicates the movement of the submarine. Doppler's effect is observed in case of light waves also. If a source of light, such as a star, is moving towards an observer on the earth, the frequency of the light increases *i.e.* the wavelength decreases and vice-versa. The phenomenon has a number of applications in the field of astronomy.

Doppler's Effect in Light: Doppler's effect is very important in light waves. If a source of light is moving towards an observer, the apparent frequency of the light will be increased (the wavelength will be decreased) and its spectral lines will be shifted towards the blue end of the spectrum. Conversely, if the source is moving away from the observer, the lines will be shifted towards the red end.

There is, however, a difference between the Doppler's effect in sound and that in light. In sound, the Doppler's effect depends upon whether the source is moving or the observer is moving, whereas in case of light it does not. This is because of the theory of relativity.

Suppose the frequency of light emitted from a source is v. We can show by the theory of relativity that if either the source or the observer is moving with a velocity v such that the distance between them is decreasing, then the apparent frequency of the source will be given by

$$\nu' = \nu \sqrt{\left(\frac{1+(v/c)}{1-(v/c)}\right)}, \qquad \text{...(i)}$$

where c is the velocity of light. Thus the frequency appears to be increased and the spectral lines will be shifted towards the blue end.

If λ and λ' are the actual and apparent wavelengths, then

$$\nu = \frac{c}{\lambda} \text{ and } \nu' = \frac{c}{\lambda'}.$$

Then, from eq. (i), we have

$$\frac{c}{\lambda'} = \frac{c}{\lambda} \sqrt{\left(\frac{1+(v/c)}{1-(v/c)}\right)}$$

or

$$\frac{\lambda'}{\lambda} = \sqrt{\left(\frac{1-(v/c)}{1+(v/c)}\right)} = \left(1-\frac{v}{c}\right)^{1/2}\left(1+\frac{v}{c}\right)^{-1/2}$$

$$= \left(1-\frac{1}{2}\frac{v}{c}+\ldots\right)\left(1-\frac{1}{2}\frac{v}{c}+\ldots\right).$$

In practice, $v << c$, as such v/c in higher powers can be ignored.

$$\therefore \quad \frac{\lambda'}{\lambda} = \left(1-\frac{1}{2}\frac{v}{c}\right)\left(1-\frac{1}{2}\frac{v}{c}\right) = 1 - \frac{v}{c}$$

or

$$1 - \frac{\lambda'}{\lambda} = \frac{v}{c}$$

or

$$\frac{\lambda-\lambda'}{\lambda} = \frac{v}{c}$$

or $$\boxed{\frac{\Delta\lambda}{\lambda} = \frac{v}{c}}$$

where $\Delta\lambda$ is Doppler's shift in wavelength.

If as a result of relative motion between the source and the observer, the distance between them is increasing, then the apparent frequency is given by

$$\nu' = \nu \sqrt{\left(\frac{1-(v/c)}{1+(v/c)}\right)}$$

In this case the frequency appears to be decreased and the spectral lines will be shifted towards the red end. Again, we shall have

$$\frac{\Delta\lambda}{\lambda} = \frac{v}{c}$$

By measuring the shift, the velocity of the source, v, can be estimated.

This has a number of applications in Astrophysics.

(i) *Velocity of Stars:* If a star is moving either towards or away from an observer on the earth, its spectral lines will be shifted either towards the blue and or towards the red end of the spectrum. The phenomenon can be used to estimate the velocity of the star.

For this, a photograph of the spectrum of the star is taken. The spectral lines are then compared with the same lines obtained by photographing the spectrum of an element in the laboratory. If the former are displaced towards the red, the stai is receding from the earth; If displaced towards the violet, the star is approaching the earth.

(ii) *Rotation of the Sun:* The sun is rotating about its own axis. If the eastern and western edges of the sun's disc are viewed, then one edge is moving towards the earth and the other is moving away from the earth. Each edge contains absorption lines due to elements such as iron vaporised in the sun, and also some absorption lines due to oxygen in the earth's atmosphere.

Therefore, if spectral photographs are taken of the east and west edges of the sun, and the two photographs are put together so that the oxygen lines coincide, then the iron lines in the two photographs are found displaced opposite to each other. By measuring this displacement the velocity of rotation of the sun can be found. Measurements show a rotational speed of about 2 km /sec.

(iii) *Spectroscopic Binaries:* Examination of the light from certain stars show a periodic doubling of the spectral lines. This indicates that each such star is, in fact, a double star revolving about one another. Therefore, while one is approaching the earth, the other is going away. These stars are known as spectroscopic binaries.

(iv) *Nabulae:* The spectral lines from certain distant nabulae appear to shift towards the red end of the spectrum. A measurement of this shift gives information about the velocities of nabulae relative to the solar system.

(v) *Saturn's Rings:* The planet saturn is surrounded by concentric rings. The nature of these rings can be established by examining the sunlight reflected by them. If the rings were solid then the velocity of the outer edge would be greater than that of the inner. Hence the light from the outer edge would show a greater Doppler shift of the Fraunhofer lines than that from the inner edge.

If, on the other hand, the rings consisted of tiny particles (satellites) moving round the saturn, the particles nearer to saturn must have a greater velocity than those farther away. Hence the light from the inner particles should show a greater shift. This is actually found to be the case, thus establishing that the rings are composed of swarms of particles.

Doppler effect has also been used for measuring the temperature of plasma.

Problems

Important Note: In order to solve the numerical problems, the following formula is enough to remember.

When both the source and the observer are moving in the direction of sound with velocities a and b respectively, the apparent frequency is given by

$$n' = n \frac{v-b}{v-a},$$

where v is the velocity of sound and n is the actual frequency.

The formulae for all other cases should be obtained by giving proper sign to the velocities a and b, according to the following rule :

> "Velocities measured in the direction of sound should be taken as positive, while those measured in the opposite direction as negative. The direction of sound is taken as the direction from the source towards the observer."

1. A person is standing near a railway track and a train moving with a speed of 36 km/h is approaching him. The apparent pitch of the whistle as heard by the person is 700 Hz. Calculate the actual frequency of the whistle. The velocity of sound is 350 m/s.

Solution: When both the source and the observer are moving in the direction of sound with velocities a and b respectively, the frequency heard by the observer is given by

$$n' = n\left(\frac{v-b}{v-a}\right)$$

where v is the velocity of sound and n is the actual frequency. Here, the source is moving in the direction of sound and the observer is stationary. Therefore, we have

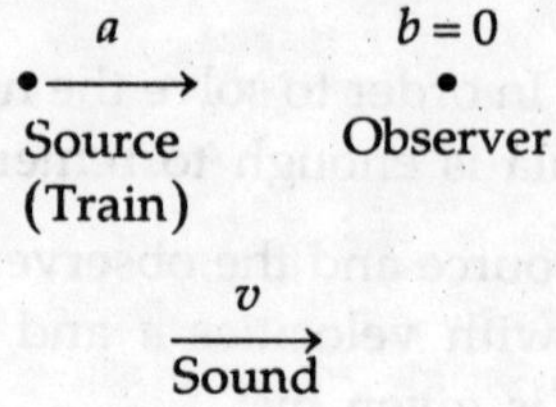

a=36 km/ h=10 m/s,

b=0, v=350 m/s, n' = 700 Hz.

$$\therefore\ 700 = n\left(\frac{350}{350-10}\right)$$

$$n = \frac{700\times 340}{350} = 680 \text{ Hz.}$$

2. *A concert goer harries down the aisle to his seat at 3 m/s. What frequency does he hear when the note A (440 Hz) is being played? The velocity of sound is 330 m/s.*

Solution. When both the source and the observer are moving in the direction of sound with velocities a and b respectively, the frequency heard by the observer is given by

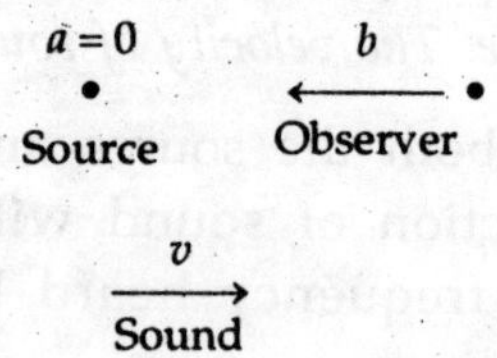

$$n' = n\left(\frac{v-b}{v-a}\right)$$

where v is the velocity of sound and n is the actual frequency.

Here, the source is stationary and the observer is moving against the sound. Therefore, we have

a=0, b= –3 m/s, v=330 m/s and n=440 Hz.

$$\therefore \quad n' = 440\left(\frac{330-(-3)}{330}\right)$$

$$= 444 \text{ Hz.}$$

3. A locomotive is passing an observer at 72 km per hour and is sounding its whistle whose frequency is 400 Hz. Determine the frequencies heard by the observer as the locomotive approaches and recedes. Velocity of sound is 332 m/sec

Solution: When both the source and the observer are moving in the direction of sound with velocities a and b respectively, the frequency heard by the observer is given by

$$n' = n\left(\frac{v-b}{v-a}\right)$$

where v is the velocity of sound and n is the actual frequency.

In this case the observer is stationary, hence b=0. The velocity of locomotive is, a=72 km/hour=20 m/sec. The velocity of sound, v=332 m/sec.

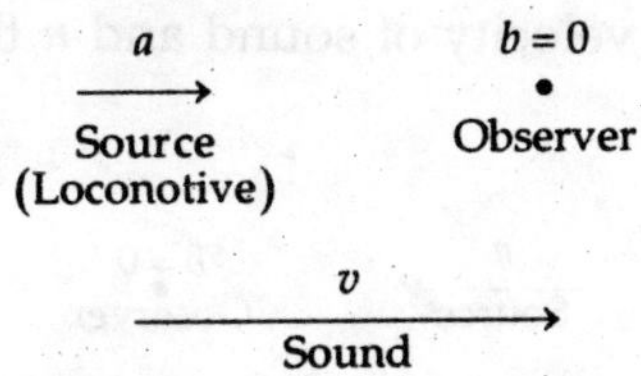

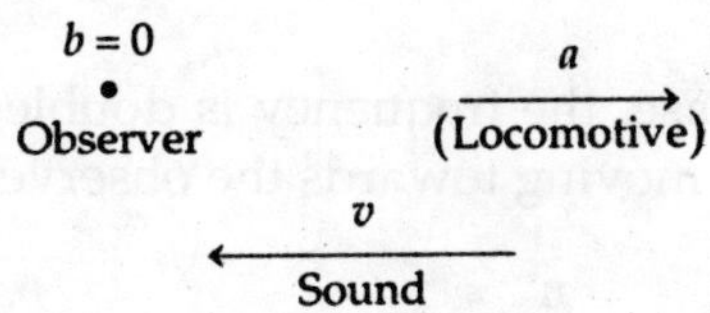

When the locomotive is approaching the observer, the frequency heard is

$$n' = n\left(\frac{v}{v-a}\right)$$

$$= 400\left(\frac{332}{3320-20}\right) = 426 \text{ Hz.}$$

After the locomotive passes the observer, a becomes negative because now the locomotive is moving in a direction opposite to that of the sound. Therefore, the frequency heard is now

$$n'' = n\left(\frac{v}{v-(-a)}\right) = 400\left(\frac{332}{3320-20}\right) = 377 \text{ Hz.}$$

4. Find the velocity of the source when the frequency appears to be (a) double, (b) half the original frequency to a stationary observer.

Solution: When both the source and the observer are moving in the direction of sound with velocities a and b respectively, the frequency appeared to the observer is given by

$$n' = n\left(\frac{v-b}{v-a}\right),$$

where v is the velocity of sound and n the true frequency of the source.

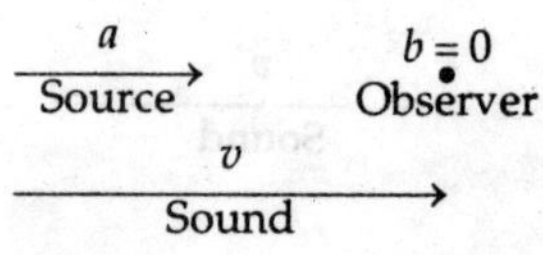

(a) In this case, the frequency is double which indicates that the source is moving towards the observer. Now, we have

$$\frac{n'}{n} = 2$$

and $b = 0$, (observer is at rest).

$$\therefore \quad 2 = \frac{v-0}{v-a}$$

$$a = v/2.$$

Hence the source is moving towards the observer with half the velocity of sound.

(b) In the case, the frequency is halved which indicates that the source is receding from the observer, a will now be negative since it is opposite to the direction of sound. We have

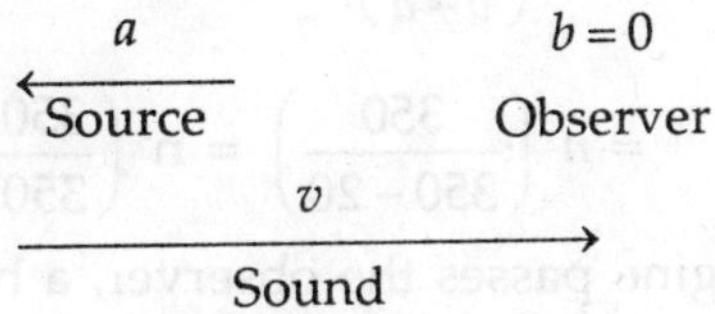

$$\frac{n'}{n} = \frac{1}{2}$$

$$\therefore \frac{1}{2} = \frac{v-0}{v-(-a)}$$

$$a = v.$$

Hence the source is receding away from the observer with the velocity of sound.

5. The pitch of the whistle of an engine moving at 72 kmh/ appears to drop by 40 Hz as it passes a stationary observer on the railway platform. Find the frequency of the whistle. The velocity of sound in air is 350 m/s.

Solution: When both the source and the observer are moving in the direction of sound with velocities a and b respectively, the frequency heard by the observer is given by

$$n' = n\left(\frac{v-b}{v-a}\right),$$

where v is the velocity of sound and n is the actual frequency

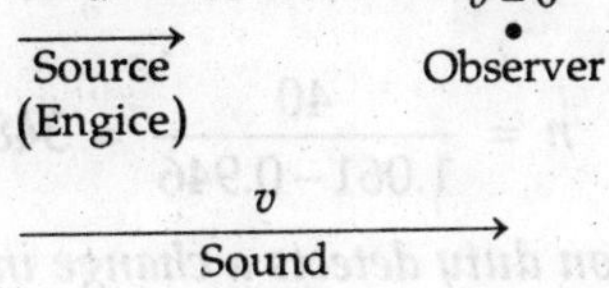

In this case the observer is stationary, hence $b = 0$.

a=72 km/h = 20 m/s and v = 350 m/s

When the engine is approaching the observer, the velocity a is positive. Hence the frequency heard is given by

$$n' = n\left(\frac{v}{v-a}\right).$$

$$= n\left(\frac{350}{350-20}\right) = \text{n}\left(\frac{350}{350}\right) = 1.061\ n.$$

After the engine passes the observer, a becomes negative because now me engine is moving in a direction opposite to that of the sound. Therefore, the frequency heard is now

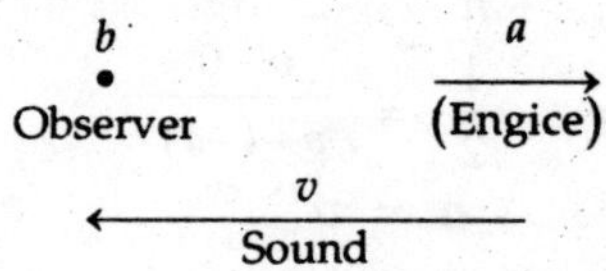

$$n'' = n'\left(\frac{v}{v-(-a)}\right)$$

$$= n\left(\frac{350}{350+20}\right)$$

$$= n\left(\frac{350}{370}\right) = 0.946n$$

As the engine passes the observer, the frequency drops by 40 Hz. Thus

$$n'—n'' = 40$$

$$1.061\ n - 0.946\ n = 40$$

$$n = \frac{40}{1.061-0.946} = 348 \text{ Hz.}$$

6. A policeman on duty detects a change in the frequency of the

horn of a car in the ratio 7:5 as the car crosses him. If the velocity of sound is 330 metres/second, calculate the speed of the car.

Solution: When both the source and the observer are moving in the direction of sound with velocities a and b respectively, the frequency heard by the observer is given by

$$n' = n\left(\frac{v-b}{v-a}\right),$$

where v is the velocity of sound and n the actual frequency.

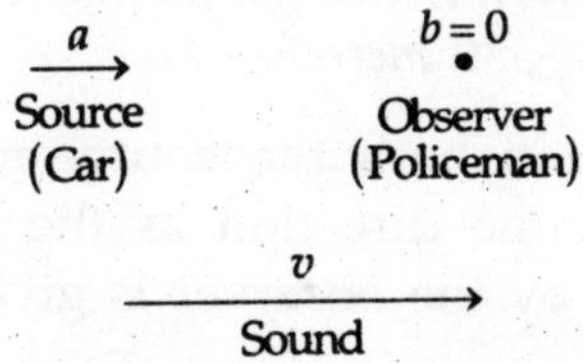

In this Case the observer (policeman) is stationary, hence b=0. When the car is approaching the policeman, the frequency heard is

$$n' = n\left(\frac{v}{v-a}\right) = n\left(\frac{330}{330-a}\right)$$

After the car passes the policeman, a becomes negative because now the car is moving opposite to the direction of sound. Therefore, now the frequency heard is

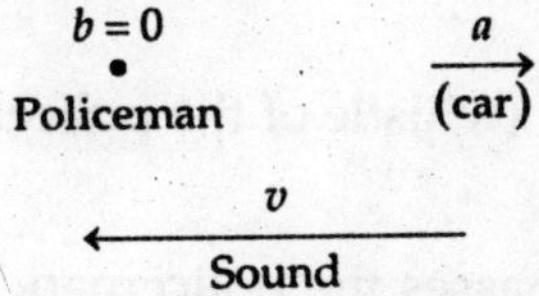

$$n'' = n\left(\frac{v}{v-(-a)}\right) = n\left(\frac{330}{330+a}\right).$$

As the car passes, there is a drop in frequency in the ratio 7 : 5. That is,

$$\frac{n'}{n''} = \frac{7}{5}$$

or $$\frac{7}{5} = \frac{330+a}{330-a}$$

whence $$a = 55 \text{ metre/sec.}$$

7. A policeman blows a whistle of frequency 300 as a car speeds past him with a velocity of 18 km/hour. Find the change in frequency as heard by the driver of the car just as he passes the policeman. (Velocity of sound -320 metre/sec.)

Solution: When both the source and the observer are moving in the same direction as the sound, the apparent frequency heard by the observer is given by

$$n' = n\left(\frac{v-b}{v-a}\right),$$

where v, a, b are the velocities of sound, source and observer respectively and n the true frequency of the source.

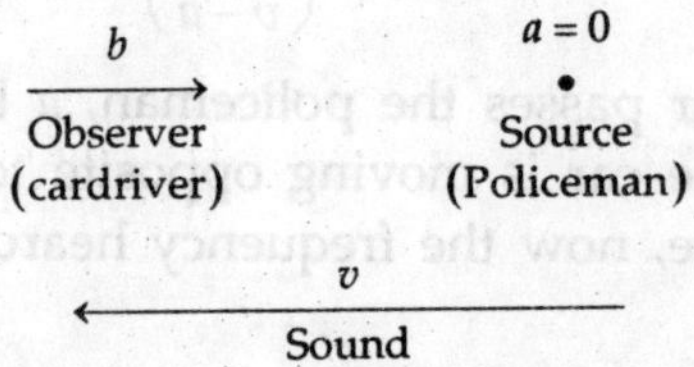

Here the source (whistle of the policeman) is stationary *i.e.* $a=0$

Before the car passes the policeman, it is moving against the direction of sound so that b is negative. Thus

b= –18 km/hour=-5 m/sec, n=300 and v=320 m/sec. Hence the frequency heard by the car driver is given by

$$n'' = 300\left[\frac{320-(-5)}{320-0}\right]$$

$$= 300\times\frac{325}{320} = 304.5 \text{ Hz.}$$

After the car passes the policeman, it is moving in the direction of sound and so b is now positive. Hence, now the frequency heard by the driver is given by

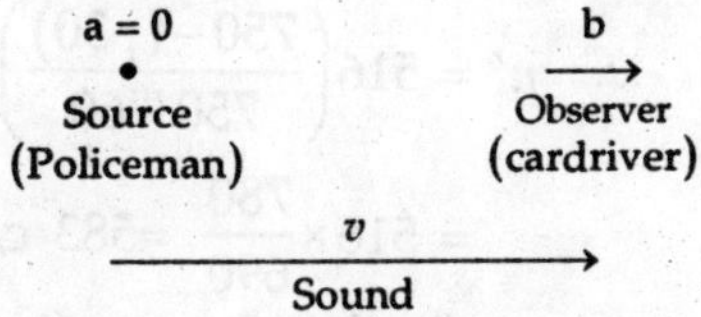

$$n'' = 300\left(\frac{320-5}{320-0}\right)$$

$$= 300\times\frac{315}{320} = 295.3 \text{ Hz.}$$

Hence the change in frequency as the car passes the policeman is

$$n' = n'' = 304.5 - 295.3 = 9.2 \text{ Hz.}$$

8. A train moving with a velocity of 60 miles per hour (mph) blows a whistle with a frequency of 516 c/s. find the note heard by a passenger in another train approaching the first at a speed of 30 mph. The velocity of sound is 750 mph.

Solution: When the source and the observer move in the same direction as the sound, the frequency of the source as heard by the observer is given by

$$n' = n\left(\frac{v-b}{v-a}\right)$$

where n is the true frequency of the note and v, a, b are the velocities of the sound, the source and the observer respectively.

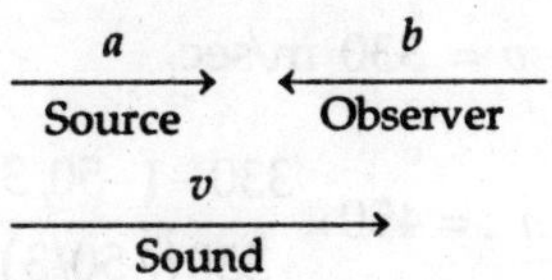

The source (first train) is moving in the direction of sound while the observer (second train) is moving opposite to the direction of sound. Thus

a=60 mph, $b= -30$ mph, v=750 mph and n=516 c/s.

$$\therefore \quad n' = 516\left(\frac{750-(-30)}{750-60}\right)$$

$$= 516\times\frac{780}{690} = 583 \text{ c/s.}$$

9. *Two engines pass each other in opposite directions, one of them blowing a whistle of frequency of 450 per second. Calculate the frequency heard on the other engine, before and after they have passed each other. Each engine moves with a velocity of 60 km/hour. Velocity of sound=330 m/sec.*

Solution: When both the source and the observer are moving in the same direction as the sound, the apparent frequency is given by

$$n' = n\left(\frac{v-b}{v-a}\right),$$

where v, a, b are the velocities of sound, source and observer respectively.

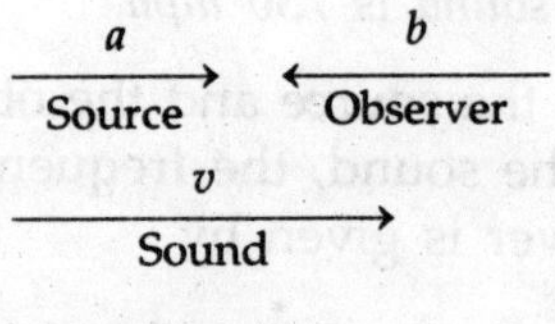

(a) Before crossing each other,

$$a = 60 \text{ km/hour} = \tfrac{50}{3} \text{ m/sec,}$$
$$b = -\tfrac{50}{3} \text{ m/sec,}$$
$$n = 450$$

and $\qquad v = 330$ m/sec.

$$\therefore \quad n' = 450\times\frac{330-(-50/3)}{330(-50/3)}$$

$$= 450 \times \frac{1040}{940}$$

$$= 498 \text{ sec}^{-1}$$

(b) After crossing each other,

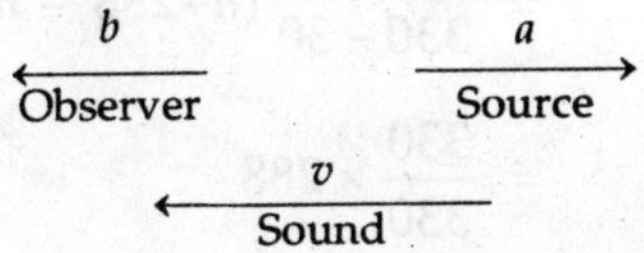

$$a = -\frac{50}{3} \text{ m/s}$$

and $$b = -\frac{50}{3} \text{ m/s}$$

$$\therefore \quad n' = 450 \times \frac{330 - (-50/3)}{330(-50/3)}$$

$$= 450 \times \frac{940}{1040}$$

$$= 407 \text{ sec.}$$

10. A motor car is fitted with two horns which differ in frequency by 288 vibrations per second. If the car sounding the horns is moving at 30 metre/second towards a person who is at rest, calculate the change in the difference of frequencies of the notes heard by the person. (Velocity of sound in air is 330 metre/second).

Solution: When the source moves with a velocity a towards the observer, the apparent frequency as heard by the observer is

$$n' = n\left(\frac{v}{v-a}\right),$$

where n is the actual frequency. Let n and n+288 be the actual frequencies of the horns. If the motor-car is moving at 30 m/sec, then a=30 m/sec, and the apparent frequencies will be

$$n \frac{330}{330-30}$$

and $$(n + 288)\frac{330}{330-30}$$

$\therefore$ the difference of frequencies of the notes heard by the observer

$$= \frac{330}{330-30}(n+288 - n)$$

$$= \frac{330}{330} \times 288$$

$$= 317 \text{ second}^{-1}.$$

11. An engine blowing a whistle of frequency 128 moves with a velocity of 72 km/hour towards a hill from which a well defined echo is heard. Calculate the frequency of the echo heard by the driver. Velocity of sound in air=330 metre/sec.

Solution: When both the source and the observer move in the direction of sound, the frequency heard is given by

$$n' = n\left(\frac{v}{v-a}\right)$$

where n is the true frequency and v, a, b are the velocities of the sound, the source and the observer respectively. The sound (echo) is reaching the observer from the hill, and can be regarded as coming from the "image" of the source formed by reflection at the hill. This is image is approaching the observer (driver) in the direction of sound so a=+72 km/hour=+20 metre/sec. The observer Is hill moving in the direction opposite to the sound coining from the hill so b=–20 metre/sec.

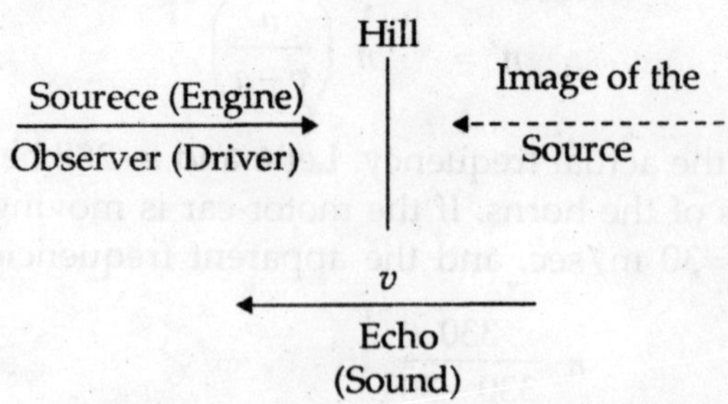

$$n' = 128\left[\frac{330-(-20)}{330-20}\right]$$

$$= 128 \times \frac{350}{310}$$

$$= 144.$$

12. *If a vibrating fork is rapidly moved towards a wall, beats are heard between the direct add reflected sounds. Calculate beat frequency if the fork makes 1000 vibrations per second and approaches the wall with a velocity of 300 cm/sec. The velocity of sound is 330 metres sec.*

Solution: When both the source and the observer are moving in the direction of sound, the frequency heard is given by

$$n' = n\left(\frac{v-b}{v-a}\right)$$

where n is the true frequency and v, a, b are the velocities of the sound, the source and the observer respectively.

Here, n=1000 vibrations/sec, v=330 m/sec, a=3 m/sec and b=0. (Observer is stationary).

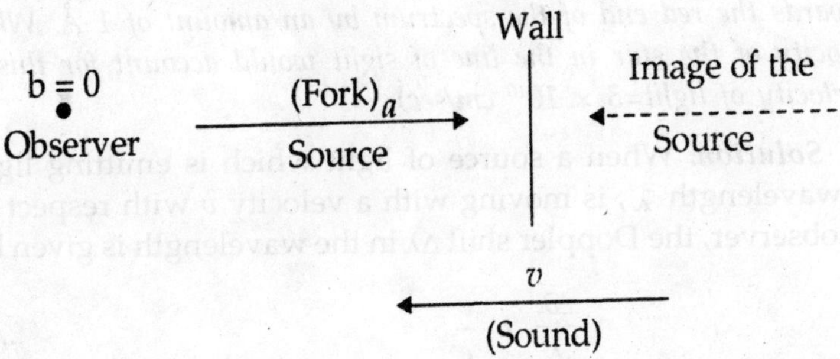

Since the source (fork) is going away from the observer, a will be negative. Therefore the sound received direct by the observer has a frequency n', given by

$$n' = 1000 \times \frac{330-0}{330-(-3)}$$

$$= 1000 \times \frac{330}{333}$$

$$= 991 \text{ per second.}$$

The reflected sound received by the observer may be regarded as coming from the "image" of the source formed by reflection at the wall. This image is approaching the observer in the direction of sound, so *a* will be positive. Hence the frequency of this sound as heard by the observer will be

$$n'' = 1000 \times \frac{330-0}{330-3}$$

$$= 1000 \times \frac{330}{327}$$

$$= 1009 \text{ per second.}$$

$$\therefore \quad \text{beat frequency} = n'' - n'$$

$$= 1009\text{–}991$$

$$= 18 \text{ sec}^{-1}.$$

13. A spectrum line of wavelength 6000 Å in the spectrum of light from a star is found to be displaced from its normal position towards the red end of the spectrum by an amount of 1 Å. What velocity of the star in the line of sight would account for this ? (Velocity of light=3 × 10^10 cm/sec).

Solution: When a source of light which is emitting light of wavelength λ, is moving with a velocity v with respect to an observer, the Doppler shift $\Delta\lambda$ in the wavelength is given by

$$\frac{\Delta\lambda}{\lambda} = \frac{v}{c} \qquad \text{...(i)}$$

where c is the velocity of light. In the given case, we have

$\Delta\lambda = 1$ Å, $\lambda = 6000$ Å and $c = 3 \times 10^{10}$ cm/second.

Substituting in eq. (i), we get

$$\frac{1}{6000} = \frac{v}{3 \times 10^{10}}$$

$$\therefore \quad v = \frac{3 \times 10^{10}}{6000}$$

$$= 50 \times 10^5 \text{ cm/second}$$

$$= 50 \text{ km/second.}$$

As the shift is towards the *red* end of the spectrum (frequency decreased), the star is moving *away* from the observer on the earth.

14. *A spectral line appearing at a wavelength of 4000 Å in the laboratory is observed at 4005 Å in the spectrum of a moving galaxy. Calculate the recession velocity of the galaxy. Velocity of light is 3×10^8 m/s.*

[Ans. 3.75×10^5 m/s.]

15. *Deduce the Doppler shift if a star emitting light of wavelength 5000 Å is approaching the observer with a velocity of 300 km/second.*

Solution: The star is approaching the observer. Therefore the Doppler shift is towards the *blue* end of the spectrum, and is given by

$$\Delta\lambda = \frac{v}{c}\lambda,$$

where v is the velocity of the star and c the velocity of light.

Here λ =5000 Å, v=300 km/second=300 × 105 cm/second and c=3×10^{10} cm/second.

$$\therefore \quad \Delta\lambda = \frac{300 \times 10^5 \text{ cm/sec}}{3 \times 10^{10} \text{ cm/sec}} \times (5000 \text{ Å})$$

$$= 5 \text{ Å.}$$

Bibliography

Anderson, J.R. Stevens D.E. and Julian P.R.: *Temporal Variations of the Tropical 40-50 Day Oscillation:* Monthly Weather Review,1984.

Bantzer, C.H. and Wallace J.M.: *Intraseasonal Variability in Tropical Mean Temperature and Precipitation,* Journal of The Atmospheric Sciences, 1996.

Blade, I. and Hartmann D.L.: *Tropical Intraseasonal Oscillations in a Simple Nonlinear Model,* Journal of the Atmospheric Sciences,1993.

Blade, Ileana: *Dynamics, Thermodynamics and Extratropical Interactions of Tropical Intraseasonal Oscillations,* University of Washington, Washington, 1994.

Chang, F.C. and Lau K.M.: *Tropical Intraseasonal Oscillation and its Prediction by the NMC Operational Model,* Journal of Climate, 1992.

Djurovic, D. and Paquet P.: *New Indications of Oscillation in Solar Activity, Atmospheric Circulation and Earth's Rotation,* Solar Physics, 1995.

Eischeid, J. Weickmann, K.M. and Khalsa S.J.S.: *The Atmospheric Angular Momentum Cycle During the Tropical Madden Julian Oscillation,* Monthly Weather Review, 1992.

Ferreira, R.N Schubert W.: *Dynamical Aspects of Twin Tropical Cyclones Associated with the Madden Julian Oscillation,* Journal of the Atmospheric Sciences, 1996.

Gao, X.H. and Stanford, J.L.: *Low Frequency Oscillations in Total Ozone Measurements*, Journal of Geophysical Research, 1990.

Gelaro, R. and Murphree T.: *Observed and Simulated Northern Hemisphere Intraseasonal Circulation Anomalies and the Influence of Model Bias,* Monthly Weather Review, 1996.

Gutzler, D.S. and Ponte, R.M.: *Oscillations in the Western Tropical Pacific,* Geophysical Research Letters, 1992.

Hartmann, D.L.; Michelsen M.L. and Klein S.A.: *Seasonal Variations of Tropical Intraseasonal Oscillations,* Journal of the Atmospheric Sciences, 1992.

Hartten, Leslie Marie: *Westerly Wind Bursts*, University of Wisconsin, Madison, 1993.

Held, I.M. Hemler, R.S. and Ramaswamy V.: *Radiative Convective Equilibrium with Explicit two Dimensional Moist Convection,* Journal of the Atmospheric Sciences, 1993.

Kessler, W.S. and McPhaden M.J.: *Forcing of Intraseasonal Kelvin Waves in the Equatorial Pacific,* Journal of Geophysical Research, 1995.

Kiladis, G.N. and Wheeler M.: *Horizontal and Vertical Structure of Observed Tropospheric Equatorial Rossby Waves,* Journal of Geophysical Research, 1995.

Kirtman, B. and Vernekar, A.: *On Wave CISK and the Evaporation Wind Feedback for the Madden Julian Oscillation,* Journal of the Atmospheric Sciences, 1993.

Lau, K.M. Sheu P.J. Schubert S. and Ledvina D.: *Evolution of Large Scale Circulation During Toga Coare Model Intercomparison and Basic Features,* Journal of Climate, 1996.

Liebmann, B.: *Organization of Convection within the Madden Julian Oscillation,* Journal of Geophysical Research, 1994.

Magana, V.: *Oscillations in Atmospheric Angular Momentum at Various Latitudes,* Journal of Geophysical Research, 1993.

Marcus, S.L and Ghil M.: *The Extratropical 40-Day Oscillation*